SpringerBriefs in Petroleum Geoscience & Engineering

Series Editors

Jebraeel Gholinezhad, School of Engineering, University of Portsmouth, Portsmouth, UK

Mark Bentley, AGR TRACS International Ltd, Aberdeen, UK

Lateef Akanji, Petroleum Engineering, University of Aberdeen, Aberdeen, UK

Khalik Mohamad Sabil, School of Energy, Geoscience, Infrastructure and Society, Heriot-Watt University, Edinburgh, UK

Susan Agar, Oil & Energy, Aramco Research Center, Houston, USA

Kenichi Soga, Department of Civil and Environmental Engineering, University of California, Berkeley, USA

A. A. Sulaimon, Department of Petroleum Engineering, Universiti Teknologi PETRONAS, Seri Iskandar, Malaysia

The SpringerBriefs series in Petroleum Geoscience & Engineering promotes and expedites the dissemination of substantive new research results, state-of-the-art subject reviews and tutorial overviews in the field of petroleum exploration, petroleum engineering and production technology. The subject focus is on upstream exploration and production, subsurface geoscience and engineering. These concise summaries (50-125 pages) will include cutting-edge research, analytical methods, advanced modelling techniques and practical applications. Coverage will extend to all theoretical and applied aspects of the field, including traditional drilling, shale-gas fracking, deepwater sedimentology, seismic exploration, pore-flow modelling and petroleum economics. Topics include but are not limited to:

- Petroleum Geology & Geophysics
- Exploration: Conventional and Unconventional
- Seismic Interpretation
- Formation Evaluation (well logging)
- Drilling and Completion
- Hydraulic Fracturing
- Geomechanics
- Reservoir Simulation and Modelling
- Flow in Porous Media: from nano- to field-scale
- Reservoir Engineering
- Production Engineering
- Well Engineering; Design, Decommissioning and Abandonment
- Petroleum Systems; Instrumentation and Control
- Flow Assurance, Mineral Scale & Hydrates
- Reservoir and Well Intervention
- Reservoir Stimulation
- Oilfield Chemistry
- Risk and Uncertainty
- Petroleum Economics and Energy Policy

Contributions to the series can be made by submitting a proposal to the responsible Springer contact, Anthony Doyle at anthony.doyle@springer.com.

Neha Saxena · Amit Kumar · Ajay Mandal

Nano Emulsions in Enhanced Oil Recovery

 Springer

Neha Saxena ⓘ
Department of Chemistry
IIMT University
Meerut, Uttar Pradesh, India

Amit Kumar
Petroleum Engineering and Geoengineering
Rajiv Gandhi Institute of Petroleum
Technology
Amethi, Uttar Pradesh, India

Ajay Mandal ⓘ
Department of Petroleum Engineering
Indian Institute of Technology (ISM)
Dhanbad, Jharkhand, India

ISSN 2509-3126   ISSN 2509-3134 (electronic)
SpringerBriefs in Petroleum Geoscience & Engineering
ISBN 978-3-031-06688-7   ISBN 978-3-031-06689-4 (eBook)
https://doi.org/10.1007/978-3-031-06689-4

This Springer imprint is published by the registered company Springer Nature Switzerland AG
The registered company address is: Gewerbestrasse 11, 6330 Cham, Switzerland

Dedicated to our parents…

Acknowledgements

We wish to express our sincerest appreciation and gratitude to the co-author of this book Dr. Ajay Mandal, Professor, Department of Petroleum Engineering, Indian Institute of Technology (Indian School of Mines), Dhanbad, for his continuous guidance, rigorous supervision, effective suggestions and the hours spent for discussing and reviewing the intricate aspects of this work.

We would like to pay our sincere thanks to IIMT University, Meerut, and Rajiv Gandhi Institute of Petroleum Technology (RGIPT), Jais, for motivating to develop a research-oriented aptitude and to utilize our knowledge in right direction.

Contents

Chapter 1
Basic Aspects of Emulsion and Nano-emulsion

1.1 Introduction

Colloids consists of a wide range of materials, and the general structure comprises of dispersing phase and dispersed phase, ranging from small molecular size to microns in length. Depending upon the stability the colloids that are thermodynamically stable are spontaneous and the ones that are metastable external energy source for their formation. Metastable colloids are generally formed with two processes (a) nucleation and growth, and (b) includes fragmentation, and in both the cases the result is formation of colloids. Colloids are known for their surface-active properties that are generally important for stabilization of the freshly formed colloids as the repulsive forces between the molecules may results in recombination of the molecules that disturbs the colloidal formation.

Emulsions are one such class of metastable colloids, formed of two immiscible components, one forming the dispersing phase and the other dispersed phase, in presence of surface-active chemicals. Emulsions are formed by shearing the two immiscible components resulting the fragmentation of the droplets. Emulsions possess all the properties of metastable colloids like Brownian movement, reversible phase transition formed due to interaction between the colloidal droplets and the irreversible transition results in liquidation of emulsion molecules [1, 2]. Due to the destruction of emulsion droplets, the volume fraction show variation that ranges from zero to one. The dense droplets of emulsions are sometimes resembling to the foams formed at air-aqueous interface, where the bulk phase is very minor. Depending upon the concentration, the emulsion possesses internal fluid dynamics and several other mechanical properties. On Dilution, the emulsion is agitated by some external source Brownian motion and behave as viscous Newtonian fluid. When emulsion is concentrated, the close packing fraction is around 64%. The dynamics of these monodispersed droplets is restricted and emulsion behave as viscoelastic solids [2, 3]. The simple emulsions are of two types: (a) oil dispersed in water (O/W); (b) the other one is water dispersed in oil (O/W) often called the inverse emulsions as shown

© The Author(s), under exclusive license to Springer Nature Switzerland AG 2022   1
N. Saxena et al., *Nano Emulsions in Enhanced Oil Recovery*,
SpringerBriefs in Petroleum Geoscience & Engineering,
https://doi.org/10.1007/978-3-031-06689-4_1

Fig. 1.1 Types of emulsions

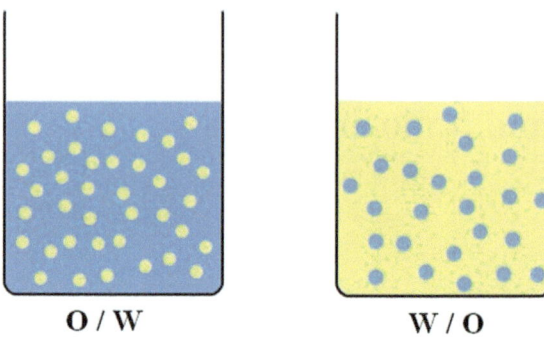

O / W   **W / O**

in Fig. 1.1. Few emulsions possess a solid dispersed phase in the dispersing medium to produce magnetic colloids [4].

Emulsions are mixture of two immiscible liquids that has a non-zero interfacial tension and may behave as lipophilic or hydrophilic fluids in presence of surfactants. The destruction of these emulsions can be processed by two mechanisms. First one is called the Ostwald ripening, in which the diffusion of the dispersed phase is through the continuous film where no rupture of the film occurs. The second one is the common one called as the coalescence that involves the rupture of the film formed at the interface of two adjacent droplets, The splitting of the two droplets is a result of formation of a thin hole between the droplets that leads to recombination of the two droplets. This is often called as the destruction of the dispersed system as the two immiscible phases can be recovered. The presence of surface-active agents enhances the lifetime of emulsion by delaying both Ostwald ripening and coalescence. Moreover, metastability of emulsion is due to the presence of surfactant at the interface. These properties of surfactant make the emulsion as good contender for industrial application.

Emulsions are used in transportation as they solubilize hydrophobic substances in liquid continuous phase. All surface treatments like printing, painting, surfacing, lubrication involves application of emulsion. Emulsions are used in food and cosmetic industry as they possess rheological properties. Emulsion have found tremendous application in medicinal field due to good interfacial properties and are stable with water as a continuous phase in a system.

Nanoemulsions are the class of kinetically stable colloidal system formed by two immiscible fluids, in which the spherical droplets of the dispersed fluid are encompassing by the other fluid forming the continuous phase. Nanoemulsion are different from the other categories of emulsions, i.e., macroemulsions and microemulsions, On the basis of its stability and droplet size. Like, traditional emulsions, nanoemulsions are also categorized as oil-in-water nanoemulsions type (O/W), in which oil droplets are dispersed in water phase, and water-in-oil nanoemulsions (W/O), where water phase is dispersed as droplets in oleic phase. There is another class of emulsion that possess smaller droplets dispersed in continuous phase, dispersed along the droplets of the dispersed phase. Such system are often called as double emulsions [5].

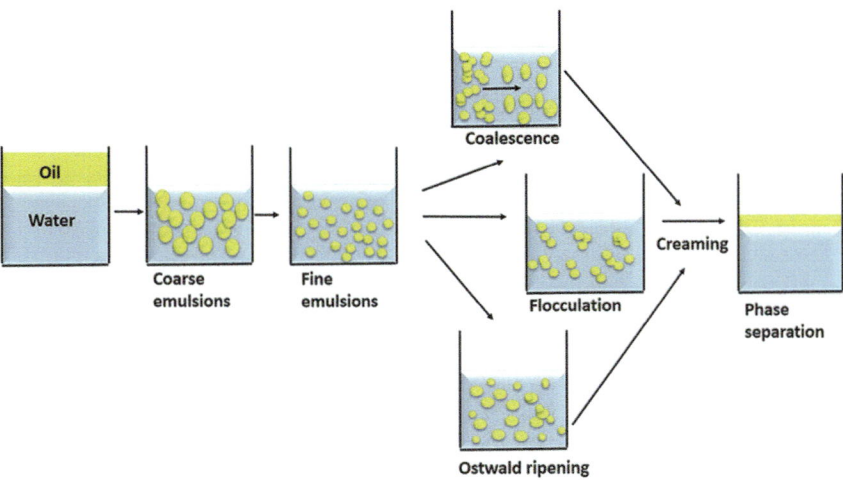

Fig. 1.2 Schematic depicting the common instability mechanisms among nanoemulsions [9]

The different physicochemical properties of the various forms of emulsions like macroemulsions, microemulsions, and nanoemulsions occur due to the variable average size distribution and the droplets stability of the respective forms of emulsions. The macroemulsions said to possess the largest average droplet size and generally thermodynamically unstable and show weak kinetic stability. Microemulsions are said to comprise of smaller droplets that are thermodynamically stable. Nanoemulsions is the different class that has average particle size in the range of < 200 nm that are kinetically stable but thermodynamically unstable. Due to kinetic stability, nanoemulsions are stable for relatively longer durations [6, 7]. However, due to thermal instability nanoemulsions are more susceptible to gravitational separation that leads to sedimentation or droplets flocculation or coalescence. The nanoemulsion are said to be prone to Ostwald ripening, resulting in phase separation [8]. Figure 1.2 shows the various mechanisms that causes the instability of a nano emulsion.

The different size of the dispersed phase in nanoemulsions also results in the difference of the physicochemical properties, such as opacity, stability, chemical behavior, and rheology, that makes nanoemulsion different from conventional emulsions. The nanoemulsion are optically clear as they show light scattering property through the droplets. The nanoemulsion stable and resistant to gravity separation as they possess weak gravitational forces. Nanoemulsions show high chemical reactivity due to larger surface area. The nanoemulsions show high volume of emulsifying layer when compared with traditional emulsions [8].

References

1. Mandal SK, Lequeux N, Rotenberg B, Tramier M, Fattaccioli J, Bibette J, Dubertret B (2005) Encapsulation of magnetic and fluorescent nanoparticles in emulsion droplets. Langmuir 21(9):4175–4179
2. Calderon FL, Stora T, Mondain Monval O, Poulin P, Bibette J (1994) Direct measurement of colloidal forces. Phys Rev Lett 72(18):2959
3. Price CP, Newman DJ (eds) (1991) Principles and practice of immunoassay. Springer
4. Lee GU, Metzger S, Natesan M, Yanavich C, Dufrêne YF (2000) Implementation of force differentiation in the immunoassay. Anal Biochem 287(2):261–271
5. Leal-Calderon F, Bibette J, Schmitt V (2007) New challenges for emulsions: biosensors, nano-reactors, and templates. In: Emulsion science. Springer, New York, pp 200–222
6. Gupta A, Burak Eral H, Alan Hatton T, Doyle PS (2016) Nanoemulsions: formation, properties and applications. Soft Matter 12(11):2826–2841
7. McClements DJ, Jafari SM (2018) General aspects of nanoemulsions and their formulation. In: Nanoemulsions. Academic Press, pp 3–20
8. McClements DJ (2011) Edible nanoemulsions: fabrication, properties, and functional performance. Soft Matter 7(6):2297–2316
9. Aswathanarayan, JB, Vittal RR (2019) Nanoemulsions and their potential applications in food industry. Front Sustain Food Syst:95

Chapter 2
Emulsion and Its Types

2.1 Emulsions

Emulsions are the combination of a heterogeneous systems consisting of at least two immiscible liquids, such as water and oil, with one of them evenly spread as small droplets throughout the other liquid phase via mechanical agitation. Emulsions are a form of colloid system that consists of a liquid and a liquid formation. The dispersed phase is made up of minute droplets, whereas the continuous phase or the dispersing phase is made up of the surrounding liquid [1, 2]. An emulsion is made up of more than just water and oil; it can also include solid particles and even gas. Due to the unfavourable interaction between the oil and water phases, an emulsion is inherently an unstable system. Although, due to the tiny drop sizes and the presence of an interfacial layer that surrounds the droplets, certain emulsions are stable [1, 3, 4]. Emulsions are thermodynamically unstable, as fusion or coalescence of droplets, the dispersed and continuous phases can revert back to distinct phases, oil and water. A mechanical force is necessary to disperse one phase into another and produce an emulsion. The emulsions generated without the addition of any surface-active material, on the other hand, may not be stable, and the emulsion phases would begin to split into layers based depending on the density differences. As a result, emulsions, on the other hand, must be stabilised by an emulsifying agent, also known as a surfactant. The more quickly coalescing droplets form the continuous phase following intense agitation of the two immiscible phases. This is generally the liquid that is present in the greatest quantity—the more droplets present, the greater the chance of collision and coalescence. As a result, emulsification may be thought of as the outcome of two concurrent processes: the disruption of bulk liquids to form small droplets and the recombination of the dispersed droplets back to the bulk liquids.

N. Saxena et al., *Nano Emulsions in Enhanced Oil Recovery*,
SpringerBriefs in Petroleum Geoscience & Engineering,
https://doi.org/10.1007/978-3-031-06689-4_2

2.2 Types of Emulsions

On the basis of the continuous phase (oleic or aqueous), emulsions are classed as (a) oil-in-water (O/W) and (b) water-in-oil (W/O). There is another class of complex emulsion called as multiple emulsion [5]. So, on emulsions are classified into 3 types on the basis of the disperse phaser and dispersing medium as shown in Fig. 2.1. The emulsion system is generally called O/W if the oleic phase is the dispersed one, while W/O if the aqueous medium forms the dispersed phase [6]. Another class called complex or multiple emulsions includes double emulsions type of system, these are complicated in their existence and their properties cannot be studied very easily. More often, these are called emulsions of emulsions [7].

2.2.1 Water-in-Oil Emulsions (W/O)

The continuous phase of a water-in-oil emulsion is generally hydrophobic materials like oil, while the dispersed phase is water [8]. The W/O type accounts for more than 95% of crude oil emulsions generated in oil fields [9]. As depicted in Fig. 2.2, the W/O emulsions include three constituents: surfactant; a solvent, and water. In the production of W/O emulsions, these mixtures play a crucial role [10].

Several studies have indicated that the most significant quality of a W/O emulsion is stability, and that these emulsions are usually stabilised themselves using natural surfactants such resin and asphaltenes [11]. The combination of W/O was divided into four states by Fingas and Fieldhouse [12]: stable, mesostable, unstable, and entrained water. In general stable emulsion appears to be brown in color and approximately contains 60–80% of water. Mesostable emulsions, such as O/W emulsions, are brownish to blackish in colour and exhibit properties that fall between stable and unstable systems. Unstable emulsions are those that split into two phases in a short period of time: water and oil and loses it emulsion properties. Finally, after a few

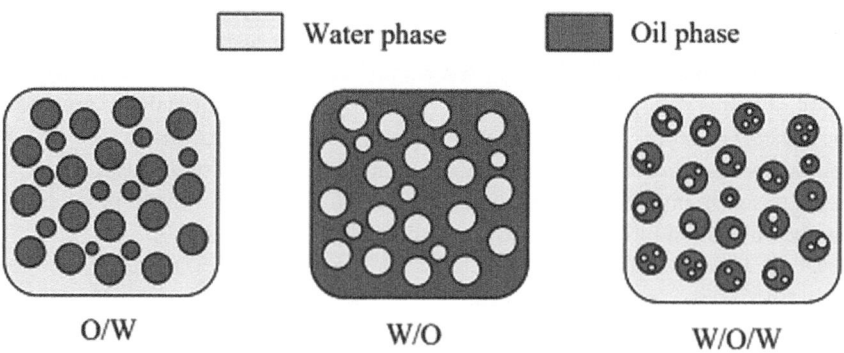

Fig. 2.1 The image depicts the classification of the emulsions [7]

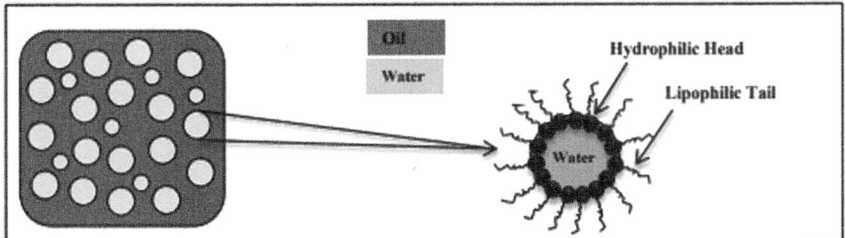

Fig. 2.2 The image depicts W/O emulsion where emulsions are stacked at the surface of dispersed phase (water) [7]

hours, the left-out water appears blackish in color and contains approximately 30–40% of water. Over the course of a week, it will eventually stabilise at roughly 10%. As stable and mesostable forms are distinguishable from the other combinations, hence they are classified as emulsions.

2.2.2 Oil-in-Water Emulsions (O/W)

As demonstrated in Fig. 2.3, an oil-in-water emulsion is one in which the oil serves as the dispersed phase and the water serves as the dispersion medium or continuous phase. If not handled appropriately, a W/O or O/W emulsion in the petroleum sector can result in significant financial losses [13]. Unlike O/W, the W/O emulsions, hence O/W emulsions are more common and hence oil in water emulsions are referred to as reversal emulsions. Pickering emulsions are another class of o/w emulsion system where stability is governed by the solid particles present between the interface of the two phases, hence lowering the surface energy and therefore have better surface active properties.

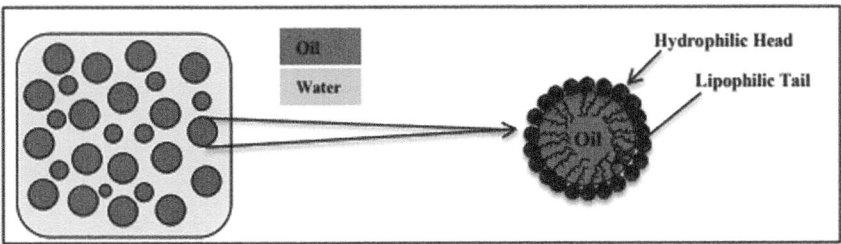

Fig. 2.3 The image depicts O/W emulsion where oleic phase is dispersed phase (oil) [7]

2.2.3 Multiple Emulsions

Multiple emulsions more often called complex emulsions are generally found as water-in-oil-in-water W/O/W (where dispersed oil globules containing smaller aqueous droplets) and oil-in-water-in-oil O/W/O type (where dispersed aqueous globules containing smaller oily dispersed droplets). Usually, these multiple emulsions are generally stabilized by the combinations of hydrophilic and hydrophobic surface-active agents. Multiple emulsions are more complicated, with extremely small droplets floating in bigger droplets that are scattered in a continuous phase. For example, water droplets entrapped in larger oil droplets are successively suspended in a continuous water phase in W/O/W emulsions. The emulsifying effect of the O/W, W/O and W/O/W type of emulsion system is linked hydrophilic lipophilic balance (HLB) [14]. HLB denotes the emulsifier's affinity for oil or water. The stronger the affinity for water, the higher the HLB scores. Emulsifiers having a greater HLB are usually better for oil-in-water emulsions. Furthermore, multiple emulsions necessitate the presence of at least two emulsifiers in the system, with one having a low-HLB (low hydrophilic lipophilic balance) and the other having a high (HLB) [15–18].

2.3 Industrial Application of Emulsions

Emulsions have found their applications in almost every area; the promising ones are petroleum industry, cosmetic industry, food and preservative industry, detergents and adhesives, inks, coatings and paint, and pharmaceutical industries. Emulsion is a type of dispersion system in which one liquid is scattered as liquid beads in another immiscible liquid. Pickering emulsion is unique in that it employs ultrafine solid particles to act as an emulsifier [19] as shown in Fig. 2.4. In 1907, Pickering discovered that solid particles may be adsorbed at the interface of oleic droplets to produce a stabilized emulsion [19]. This type of formation are referred as Pickering emulsion, where the solid particles are adsorbed and they stabilise the system are

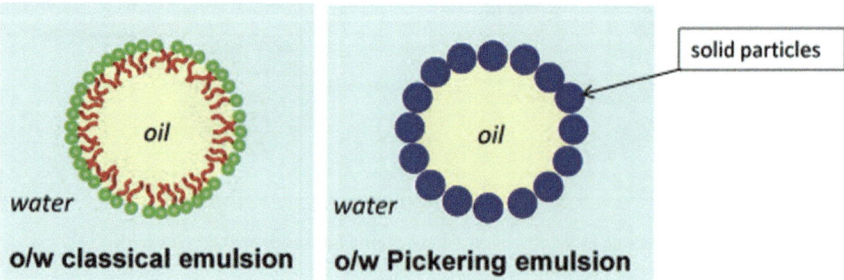

Fig. 2.4 Schematic diagram depicts the classical emulsion and the Pickering emulsion formations [21]

known as Pickering particles. Pickering emulsion differs from typical emulsion in that the emulsifier is made up of solid particles. Pickering emulsion also offers greater stability, less emulsifier, recyclability, and environmental friendliness as compared to ordinary emulsion [20].

2.3.1 Application in Petroleum Industry

Pickering emulsion are emerging as promising emulsion system for application in petroleum and gas industry. Their application in oil and gas industry includes chemical enhanced oil recovery, sand control, oil spill treatment and drilling fluid system [22, 23]. In addition, the smart Pickering emulsion systems show great promise in other areas of oil recovery and transportation. Oil spills cause severe emulsification, which has serious environmental repercussions and makes oil collection and cleanup more difficult. Depending on the process of emulsification, movement, and ripening of Pickering emulsion systems, numerous studies have been conducted using Pickering particles as an intermediate to prepare smart Pickering emulsion systems that can be controllably stabilised, induced, ripened, and help in directional transport of small sized oleic droplets in the sea under the effect of magnetic fields to identify the oil spill [24].

2.3.2 Application in Food Industry

Emulsions and emulsifiers are used extensively in various sectors, including the food industry. Emulsions include products like soft beverages, milk, cream, salad dressings, mayonnaise, soups, sauces, dips, butter, and margarine. Oil-in-water (O/W) emulsions were traditionally created by mixing oleic and aqueous phase in the presence emulsifiers [25, 26]. These are adsorbed at the surfaces of newly produced oleic droplets during homogenization, lowering the IFT (interfacial tension) and allowing for additional droplet disintegration. Furthermore, they form a carapace barrier surrounding the droplets, which aids in preventing the agglomeration aggregation formed by repulsion forces.

The commonest emulsifiers in the food industry are polysaccharides, phospholipids, amphiphilic proteins, and several surface active agents [26, 27]. Emulsifiers differ significantly in their capacity to produce tiny droplets of oil all through homogenization and avoid droplet agglomeration under various ambient conditions such as pH, heat, ionic strength, low temperature [26, 28]. Easily availability and use, cost efefctiveness and compatibility with other substances, and "label friendliness" [29] are additional factors that makes emulsifier is perfect for food industry. The best one emulsifier for a given food product is determined by the kind and composition of other ingredients it includes, the method of preparation, and the environmental conditions it encounters during manufacturing, storage, and use.

2.3.3   Application in Cosmetic Industry

Cosmetic goods come in a variety of physicochemical forms, depending on their intended function and the substances they include. They come in multiphase systems like emulsions (creams, lotions, and hair conditioners), suspensions (toothpastes, face masks, and make-up cosmetics), and foams, as well as single-phase systems like solutions (tonics, toilet waters). Color cosmetics, such as face powders or make-up foundations, can take the form of powdered solids mixes.

When it comes to the effect of cosmetics on the skin, emulsions can be classified as cleaning, moisturising, or protecting cosmetics. Day creams (quick-breaking, instant-absorbing products with no greasy after feel) and night creams (late-breaking, instant-absorbing products with no greasy after feel) are two types of creams that can be used during the day (generally with higher levels of occlusive emollients, containing nutrients). Cosmetic emulsions are divided into four categories: face creams (including eye creams), hand creams, foot creams, and body emulsions [30].

2.3.4   Application in Pharmaceutical Industry

Emulsions are widely employed in a variety of sectors. They are employed in the pharmaceutical industry that makes medications more appetizing, to increase the efficacy of active components by managing dose, and to improve the aesthetics of topical therapies like ointments. Because of their low innocuity, and property for the direct injection, and congruity with other medicinal components, nonionic forms of emulsions are the most common.

2.3.5   Application in Paint and Ink Industry

Emulsions are used in a lot of paints and inks. These products can be either real liquid-in-liquid or in the form of dispersions. The dispersion forms resemble to emulsions, however the dispersed phase is often made up of finely split solid particles. The surfactant technique that is used to make emulsions is also employed to make pigment dispersions for printing and paint industry. These forms are made to dry rapidly and produce waterproof coatings while leaving the colour unaffected. Emulsions outperform solvent-based systems in this aspect due to their lower odour and flammability.

References

1. Chrisman E, Lima V, Menechini P (2012) Crude oil emulsion-composition stability and characterization. InTech 3rd ed InTech Janeza Trdine 9(51000):1–240
2. Ichikawa T (2007) Electrical demulsification of oil-in-water emulsion. Colloids Surf A Physicochem Eng Aspects 302(1–3):581–586
3. Pal R (2002) Novel shear modulus equations for concentrated emulsions of two immiscible elastic liquids with interfacial tension. J Nonnewton Fluid Mech 105(1):21–33
4. Zhu H, Guo Z (2016) Understanding the separations of oil/water mixtures from immiscible to emulsions on super-wettable surfaces. J Bionic Eng 13(1):1–29
5. Hoshyargar V, Marjani A, Fadaei F, Shirazian S (2012) Prediction of flow behavior of crude oil-in-water emulsion through the pipe by using rheological properties. Orient J Chem 28(1):109
6. Auflem IH (2002) Influence of asphaltene aggregation and pressure on crude oil emulsion stability
7. Nour AH (2018) Emulsion types, stability mechanisms and rheology: a review. Int J Innov Res Sci Stud (IJIRSS) 1(1)
8. Foley JT, Forest P, Rogers RH (1966) United States patent 0
9. Ali MF, Alqam MH (2000) The role of asphaltenes, resins and other solids in the stabilization of water in oil emulsions and its effects on oil production in Saudi oil fields. Fuel 79(11):1309–1316
10. Kale SN, Deore SL (2017) Emulsion micro emulsion and nano emulsion: a review. Syst Rev Pharm 8(1):39
11. Fridjonsson EO, Graham BF, Akhfash M, May EF, Johns ML (2014) Optimized droplet sizing of water-in-crude oil emulsions using nuclear magnetic resonance. Energy Fuels 28(3):1756–1764
12. Fingas M, Fieldhouse B (2004) Formation of water-in-oil emulsions and application to oil spill modelling. J Hazard Mater 107(1–2):37–50
13. Abdel-Raouf ME (2012) Factors affecting the stability of crude oil emulsions. In: Crude oil emulsions-composition, stability and characterization. Intech, Croatia, pp 183–204
14. Martin MJ, Trujillo LA, Carmen Garcia M, Carmen Alfaro M, Muñoz J (2018) Effect of emulsifier HLB and stabilizer addition on the physical stability of thyme essential oil emulsions. J Dispersion Sci Technol 39(11):1627–1634
15. Yaqoob Khan A, Talegaonkar S, Iqbal Z, Ahmed FJ, Khar RK (2006) Multiple emulsions: an overview. Curr Drug Deliv 3(4):429–443
16. Muschiolik G (2007) Multiple emulsions for food use. Curr Opin Colloid Interface Sci 12(4–5):213–220
17. Pal R (2011) Rheology of simple and multiple emulsions. Curr Opin Colloid Interface Sci 16(1):41–60
18. Jiao J, Burgess DJ (2003) Rheology and stability of water-in-oil-in-water multiple emulsions containing span 83 and tween 80. AAPS PharmSci 5(1):62–73
19. Pickering SU, Emulsions J (1907) Chem. Soc 91:2001–2021
20. Yang FW, Qiang JL, Sun DLC (2009) Research progress on Pickering emulsions. Progr Chem 21(0708):1418
21. Chevalier Y, Bolzinger M-A (2013) Emulsions stabilized with solid nanoparticles: Pickering emulsions. Colloids Surf A Physicochem Eng Aspects 439:23–34
22. Sharma T, Velmurugan N, Patel P, Chon BH, Sangwai JS (2015) Use of oil-in-water Pickering emulsion stabilized by nanoparticles in combination with polymer flood for enhanced oil recovery. Pet Sci Technol 33(17–18):1595–1604
23. AfzaliTabar M, Alaei M, Ranjineh Khojasteh R, Motiee F, Rashidi AM (2017) Preference of multi-walled carbon nanotube (MWCNT) to single-walled carbon nanotube (SWCNT) and activated carbon for preparing silica nanohybrid pickering emulsion for chemical enhanced oil recovery (C-EOR). J Solid State Chem 245:164–173
24. Song Y, Zhou J, Fan J-B, Zhai W, Meng J, Wang S (2018) Hydrophilic/oleophilic magnetic Janus particles for the rapid and efficient oil–water separation. Adv Func Mater 28(32):1802493

25. Friberg S, Larsson K, Sjoblom J (eds) (2003) Food emulsions. CRC Press
26. McClements DJ (2004) Food emulsions: principles, practices, and techniques. CRC Press
27. Charalambous G, Doxastakis G (1989) Food emulsifiers: chemistry, technology, functional properties and applications. No. Sirsi, i9780444873064
28. Garti N, Reichman D (1993) Hydrocolloids as food emulsifiers and stabilizers. Food Struct 12(4):3
29. Stauffer CE (1999) Dressing and sauces. In: Emulsifiers. Eagan Press, St. Paul, MN
30. Sikora E (2019) Cosmetic emulsions: monograph

Chapter 3
Nano-emulsions

Nano-emulsions are the homogenous solution forms by two immiscible liquids with nanometer size ranging from 20 to 500 nm and is stabilized by the presence of surface-active agents. There is a lot of ambiguity in the literature about the accurate definition of nanoemulsions, these are often referred as sub-micro emulsion (SME) or mini-emulsion and are frequently muddled with thermodynamically stable and spontaneously formed microemulsions [1–3]. Generally, nano is equal to 10^{-9} and micron is equal to 10^{-6}; and in nano emulsions, the resulted droplets are smaller than in microemulsions. In reality from the experimental analysis, however, the droplet size of micro-emulsions is less than those of nano-emulsions. This perplexing phrase stems from the evolution of colloidal science throughout time. The word "microemulsion" was first used in 1961 [4], whereas "nano-emulsion" was first used in 1996 [5]. As a result, microemulsions were well-known among researchers long before nano-emulsions were introduced. The production techniques, droplet sizes, and stability behaviour of macro-emulsions, also known as classical emulsions, microemulsions, and nano-emulsions are the main differences. When given enough time, phase separation happens in macro-emulsions and nano-emulsions, indicating that both are thermodynamically unstable. Nano-emulsions, on the other hand, stay stable for extended periods of time (kinetically stable) because of their oil droplet size is of nano range, and the meta-stability of nanoemulsions has no bearing on their proclivity to a stable state [3].

Micro-emulsions are generally thermodynamically stable at equilibrium condition but are susceptible to temperature variation and composition. Figure 3.1 depicts the production techniques, droplet sizes, and stability behaviour of macro-emulsions, micro-emulsions, and nano-emulsions.

© The Author(s), under exclusive license to Springer Nature Switzerland AG 2022
N. Saxena et al., *Nano Emulsions in Enhanced Oil Recovery*,
SpringerBriefs in Petroleum Geoscience & Engineering,
https://doi.org/10.1007/978-3-031-06689-4_3

Parameters	Macro-emulsion	Micro-emulsion	Nanoemulsion
	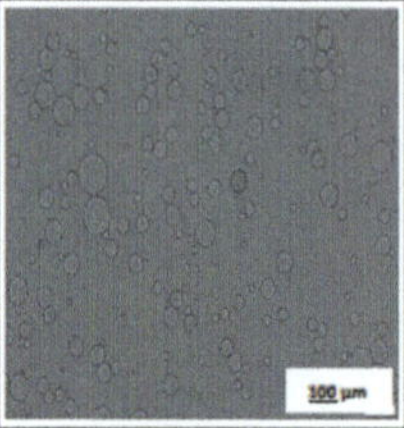	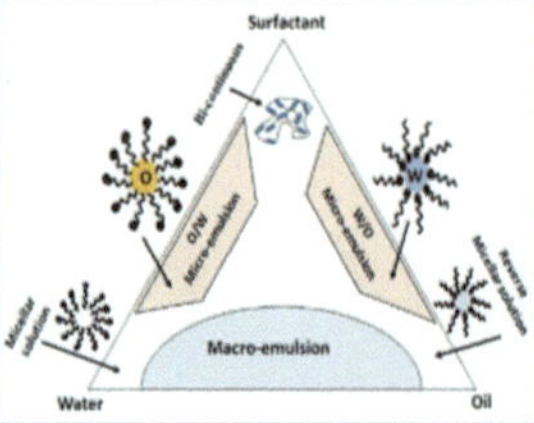	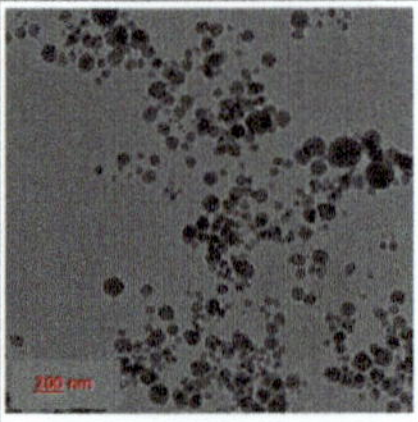
Preparation method	Low and high energy methods	Low energy methods	Low and high energy methods
Surfactant concentration	High	High	Low
Droplet size	1-100 μm	100 nm-1 μm	10-500 nm
Polydispersity	> 30-40 %	Low < 10 %	Low < 10-20 %
Droplet shape	Spherical	Spherical (direct or reverse micelles), Cylindrical (rod or reverse micelles), Lamellar structures or Bi-continuous (sponge like)	Spherical
Interfacial tension	High	Ultra-low	Ultra-low (< 10 dyn/cm)
Stability	Weakly kinetically stable and thermodynamically un-stable	Thermodynamically stable	Kinetically stable
Visual appearance	Turbid	Transparent/Clear	Transparent/Translucent
Preparation cost	Low and high cost	Low cost	High cost

Fig. 3.1 Comparison between macro-emulsion, micro-emulsion and nanoemulsion [6]

3.1   Types of Nano-emulsions

Depending upon the composition of oil and water percentages; nano-emulsion can be classified into three major types:

(a)   Oil in water (O/W) nano-emulsions: In this type of nano-emulsion system the oleic droplets are found to be dispersed in continuous aqueous phase.

(b)   Water in oil (W/O) nano-emulsions: In this type of nano-emulsion system the water droplets are found to be dispersed in continuous oleic phase.

(c)   Bi-continuous nano-emulsions: This is a complex type of system where the oleic and aqueous phases are mixed throughout the system.

The interface of these nano-emulsion system has been stabilized by using a suitable amalgamation of surfactants/co-surfactants in specific ratio as shown in Fig. 3.2.

Depending upon the type of surfactants employed the O/W nano-emulsions system are further classified into three types: (a) Neutral O/W nanoemulsions: where neural surfactants are used; (b) anionic O/W nanoemulsion: where anionic surface

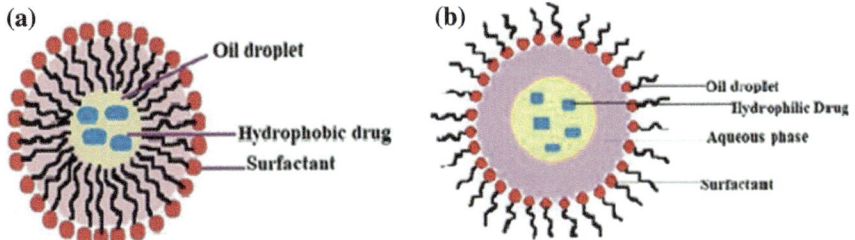

Fig. 3.2 The image depicts the O/W nano-emulsion and W/O nano-emulsion [7]

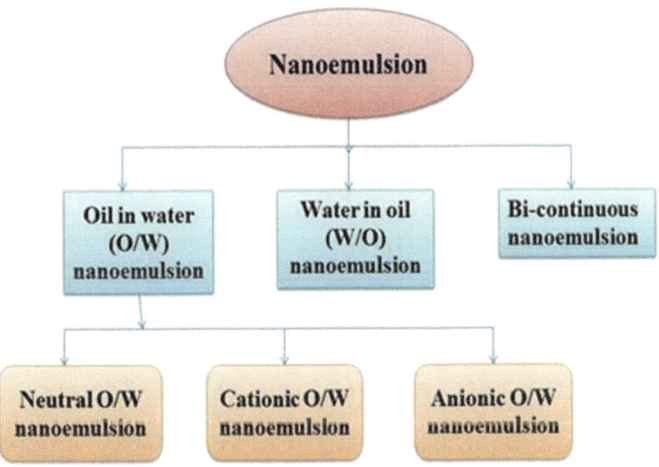

Fig. 3.3 The image depicts the classification of nano-emulsion [7]

active agents are used; (c) cationic O/W nanoemulsion: where cationic surface active agents are used. The following classification is depicted in Fig. 3.3.

3.2 Advantages of Nano-emulsions Over Emulsions

Nano-emulsions due to their tiny droplet size have various benefits over traditional emulsions. Strong physical stability against droplet agglomeration and gravitational separation, high optical clarity, and improved bioavailability of encapsulated compounds are only a few of the benefits that make them appropriate for culinary applications. Because the small size of droplet ensures a weaker scattering of light by final products, higher transparency is a very favourable and alluring feature of nanoemulsions in the beverage industry [8]. Therefore, the addition of functional ingredients to a nanoemulsion system has no discernible effect on the appearance of a final beverage product.

The physico-chemical and structural features of nanoparticles generated are adjusted according to the intended use while delivering nutraceuticals, vitamins, medicines, antimicrobials, colours and flavors [9, 10]. Since they possess superior emulsifying capabilities, good binding capacity for hydrophobic bioactive chemicals, and excellent gelation qualities, food protein stabilised nano-emulsions have been investigated as delivery vehicles. Dairy-based proteins, like whey proteins and casein, are the most frequently utilised in emulsion stabilisation. Plant proteins (such as soy, pea, lentil plant, and canola proteins) are underappreciated in the form of emulsifiers. Plant proteins, on the other hand, have sparked renewed study in recent years [11]. Many studies have found that nanoemulsions offer superior characteristics than traditional emulsions formulated for delivery methods [12, 13].

3.3   Application of Nano-emulsions

Nanoemulsions exhibit a number of distinct characteristics, that includes tiny droplet size, high stability, transparency, and adjustable rheology. Nanoemulsions are an appealing possibility for use in the food, cosmetics, pharmaceutical, and drug delivery sectors because of their distinct characteristics.

3.3.1   Applications of Nano-emulsions in Drug Delivery

Nanoemulsions have found their application in the majority of drug delivery methods, including topical, ophthalmic, internasal, intravenous, and oral. The following applications take advantage of nanoemulsions' lyphophilic nature to solvate water-insoluble medicines, as well as their adjustable charge and rheology to make solutions which can be conveniently given to patients. However, skin shields us from the outside world, it also serves as a transference barrier for medications administered via skin. Topical medications that are formulated with nanoemulsions has distinct advantages because of the dispersed phase of O/W nanoemulsions allows the increased lipophilic drug solubility in the oleic phase, while the continuous phase offers a skin-friendly and mild atmosphere that dissolves biopolymers like alginate to adjust the appearance, texture, and formulation rheology. Several research groups have looked at the potential of nanoemulsions for topical drug delivery [14, 15].

3.3.2   Applications of Nano-emulsions in Food Industry

Nano-emulsions are utilised in the food and beverage industry to create smart meals with nutrients that are difficult to include because to their poor water solubility; one such example is -carotene, a pigment that gives colour to plants like carrots and has essential health advantages. The size and stability of nano-emulsions containing—carotene have been investigated by researchers in relation to temperature, pH, and surfactant type [16]. Curcumin, an anti-inflammatory substance, is another ingredient that's frequently found in nano emulsions. The mouse ear inflammatory model was used by Wang and his colleagues to investigate the anti-inflammatory response of this nanoemulsions found in curcumin [17]. Due to the simple lipid digestion stage in nanoemulsions, the researchers found that curcumin nano emulsions formed in the oleic phase can be easily digested than curcumin taken directly [15].

3.3.3 Application of Nano-emulsions as Building Blocks

Nano-emulsions due to their tiny size and large surface area, have been utilised as building blocks for the fabrication of more complicated materials, allowing for the simple decorating of a surface of liquid with functional pieces such as trendy macro-molecules. Emulsion polymerization, in which hydrophobic monomers trapped in droplets are polymerized to generate polymeric particles, and is possibly the most well-known example in polymer synthesis. In polymer synthesis, nano-emulsions (known as mini-emulsions) have been widely used [15].

3.3.4 Application of Nano-emulsions in Pharmaceuticals Industry

Over the last few years, the application of nanotechnology in drugs and medicine has increased. The phrase "NANOPHARMACEUTICALS" refers to medications produced using nanotechnology. Nanoemulsions, nanosuspensions, nanotubes (arrangement of nanoscale C60 atoms organized in a long thin cylindrical structure), nanospheres (drug nanoparticles in polymer matrix), nanoshells (concentric spherical nanoparticles comprising of a metal shell with dielectric core), lipid nanoparticles (encloser of lipid monolayer over a solid lipid core), nanocapsules (encapsulated drug nanoparticle). Nanoemulsions are a type of dispersed particle employed in pharmaceutical and biomedical aids and vehicles, and they have a bright future in cosmetics, diagnostics, medication treatments, and biotechnologies [18].

3.3.5 Application of Nano-emulsions in Oil and Gas Industry

Nanoemulsions are used in petroleum industry in recent years, but chemicals have been used for a long time, such as surfactants, which were applied in oil fields in well bore clean up systems during drilling and for enhanced oil recovery (EOR), even before 1980s' drop in oil prices. With an ever-increasing population and an ever-increasing need for energy, the petroleum industry required technical innovation while also considering environmental implications. The traditional chemicals (surface active agents) employed for oil production and well bore cleaning were found to be effective, but were susceptible in damaging the formation due to poor selection of surfactant and a lot of turbulent flow and large volumes was required, which resulted in lower oil production rates and higher operational costs. As a result, nanotechnology presents a solution in the shape of nanoemulsions to reduce the problems of traditional chemicals (surface active agents) in oil and gas sector operations [6].

3.4 Challenges for Industrial Application of Nano-emulsions

Several research have been conducted in recent years to determine the benefits of encapsulating lipophilic and functional chemicals in nanoemulsions for diverse industrial applications. In vitro studies have revealed that nano-emulsification protects and increases the bioavailability of bioactive substances. However, there are few studies that demonstrate the real health advantages of application of nanoemulsions in foods, as well as their consumption, labelling, and public's opinion.

Similarly, research has been conducted to assess the use of a high or low energy method to construct nanoemulsions, with the goal of optimising processing constraints and materials used in nanoemulsion creation. However, little research has been carried out to reduce the cost of producing nanoemulsions because their formulation and application for fortification and packaging of foods demands more energy and equipment. In the same way, the hazards of using tailored nanoemulsions in food are unknown. After digestion, the toxicological consequences and biological destiny of nanoparticles have yet to be determined [19].

References

1. El-Aasser MS, David Sudol E (2004) Miniemulsions: overview of research and applications. JCT Res 1(1):21–32
2. McClements DJ (2012) Nanoemulsions versus microemulsions: terminology, differences, and similarities. Soft Matter 8(6):1719–1729
3. Delmas T, Piraux H, Couffin A-C, Texier I, Vinet F, Poulin P, Cates ME, Bibette J (2011) How to prepare and stabilize very small nanoemulsions. Langmuir 27(5):1683–1692
4. Schulman JH, Montagne JB (1961) Formation of microemulsions by amino alkyl alcohols. Ann New York Acad Sci 92(2):366–371
5. Calvo P, Vila-Jato JL, Alonso MJ (1996) Comparative in vitro evaluation of several colloidal systems, nanoparticles, nanocapsules, and nanoemulsions, as ocular drug carriers. J Pharm Sci 85(5):530–536

6. Kumar N, Verma A, Mandal A (2021) Formation, characteristics and oil industry applications of nanoemulsions: a review. J Petrol Sci Eng 206:109042
7. Halnor VV, Pande VV, Borawake DD, Nagare HS (2018) Nanoemulsion: a novel platform for drug delivery system. J Mat Sci Nanotechol 6(1):104
8. Wang T, Soyama S, Luo Y (2016) Development of a novel functional drink from all natural ingredients using nanotechnology. LWT 73:458–466
9. Esmaeili A, Gholami M (2015) Optimization and preparation of nanocapsules for food applications using two methodologies. Food Chem 179:26–34
10. Mehmood T (2015) Optimization of the canola oil based vitamin E nanoemulsions stabilized by food grade mixed surfactants using response surface methodology. Food Chem 183:1–7
11. Yerramilli M, Longmore N, Ghosh S (2017) Improved stabilization of nanoemulsions by partial replacement of sodium caseinate with pea protein isolate. Food Hydrocolloids 64:99–111
12. Mehrnia M-A, Jafari S-M, Makhmal-Zadeh BS, Maghsoudlou Y (2016) Crocin loaded nano-emulsions: factors affecting emulsion properties in spontaneous emulsification. Int J Biol Macromol 84:261–267
13. Luo X, Zhou Y, Bai L, Liu F, Deng Y, McClements DJ (2017) Fabrication of β-carotene nanoemulsion-based delivery systems using dual-channel microfluidization: physical and chemical stability. J Colloid Interface Sci 490:328–335
14. Baboota S, Shakeel F, Ahuja A, Ali J, Shafiq S (2007) Design, development and evaluation of novel nanoemulsion formulations for transdermal potential of celecoxib. Acta Pharm 57(3):315
15. Gupta A, Burak Eral H, Alan Hatton T, Doyle PS (2016) Nanoemulsions: formation, properties and applications. Soft Matter 12(11):2826–2841
16. Yuan Y, Gao Y, Mao L, Zhao J (2008) Optimisation of conditions for the preparation of β-carotene nanoemulsions using response surface methodology. Food Chem 107(3):1300–1306
17. Wang X, Jiang Y, Wang Y-W, Huang M-T, Ho C-T, Huang Q (2008) Enhancing anti-inflammation activity of curcumin through O/W nanoemulsions. Food Chem 108(2):419–424
18. Shah P, Bhalodia D, Shelat P (2010) Nanoemulsion: a pharmaceutical review. Syst Rev Pharm 1(1)
19. Aswathanarayan JB, Vittal RR (2019) Nanoemulsions and their potential applications in food industry. Front Sustain Food Syst:95

Chapter 4
Nanoemulsions in Petroleum Industry

Petroleum, also called as crude oil, is a raw material for various types of transportation fuels, fuels for electricity and heat generation, and feedstocks for chemicals and plastics. It is a type of fossil fuel formed from the buried remains of animals and plant that existed in the marine environment. These buried biological matters have been under heat and pressure, covered by the layers of rock for over millions of years, causing it to convert to crude oil or petroleum [1].

The petroleum industry, also known as oil and natural gas industry, is a major part of energy market and plays a dominant role in the global economy as the world's prime source of fuel. The petroleum industry can be sub-divided into three sectors: upstream sector, midstream sector, and downstream sector. Upstream sector companies handle the exploration and production. The companies in midstream companies deal with transportation and storage of petroleum. Downstream companies focus on the refining and marketing of the petroleum and petroleum products. Large companies such as Chevron and Royal Dutch Shell have integrated company profile and manages all three sectors of the petroleum industry.

The exploration and production activities carried out by the upstream companies includes the processes involved in identification of potential subsurface locations for oil and gas drilling and extraction. These potential locations can be either onshore or offshore which greatly controls the cost of the operation of this capital-intensive industry. Drilling and recovery of oil and gas requires expensive equipment and highly skilled workforce [2]. After the unrefined crude oil is brought to the surface, it is sent to refineries via tankers, pipelines, barges, trucks or tugboats. The downstream companies refines the crude oil and processes natural gas into several usable products, such as, gasoline, jet fuel, diesel, asphalt and road oil, heavy fuel oil, petrochemical feedstocks, and lubricants [3].

Nanoemulsions, which are kinetically stable colloidal system of oil and water have several applications in petroleum industry, due to their drop size, stability and physicochemical properties. Some applications of nanoemulsions in the different operations of petroleum industry have been emphasized as follows.

© The Author(s), under exclusive license to Springer Nature Switzerland AG 2022
N. Saxena et al., *Nano Emulsions in Enhanced Oil Recovery*,
SpringerBriefs in Petroleum Geoscience & Engineering,
https://doi.org/10.1007/978-3-031-06689-4_4

4.1   Wellbore Cleaning

A wellbore acts as a faucet to the subsurface hydrocarbon reservoir for recovering the subsurface fluids, or as an injection point for injecting required fluids into the reservoir. These connections between the surface facilities and the subsurface reservoir rock are formed by drilling through the earth, where a drill bit scrapes, gouges, or chisels the rock formation by rotary motion provided by a drilling string. These wells are filled with drilling mud during the process to lubricate the bit, move the drill cuttings to the surface, and maintain the wellbore integrity. The drilling mud, also known as, drilling fluid, can be a water based mud (WBM), or an oil based mud (OBM). Irrespective of the type of drilling mud used for drilling, there is requirement of efficient wellbore cleaning, filter cake removal, and clean-up of formation, prior to cementing and casing process. The presence of filter cake leads to poor bonding between formation and cement, which can problems such as poor zone isolation and annular gas migration [4]. Wellbore cleaning is also required in the cased zones as presence of mud contaminants and formation debris in casing walls can cause harm to casings, drill strings, and completion equipment. A well drilled using OBM usually leads to higher deposition of oil-wet solid particles on wellbore surface and casing wall in comparison to drilling using WBM [5]. Thus, OBM have least compatibility with cement slurry. A properly cleaned wellbore is vital and the lack of it can cause damage to the completion equipment, leading to loss of production at later stages and also reduce the formation productivity. The accumulation of drill cuttings, debris and rock particles in the wellbore can lead to stuck pipe and failure of drilling tolls, thus leading to significant downtime of rig. A wellbore cleaned properly can protect the well and give assurance of wellbore integrity. The wellbore cleaning becomes more difficult and important in deviated and horizontal wells, as well as, in offshore locations.

The wellbore cleaning is done by mechanical or chemical methods. The mechanical methods of wellbore cleaning involve running in of tools such as, scrapers, brusher, and magnet casing [6]. These tools remove the filter cake formed on the wellbore surface during the drilling process by mechanical scratching of the deposited cake. Chemical methods of wellbore cleaning involve usage of chemical solutions that break filter cake, alter wettability of casing walls, and reduce the viscosity of OBMs. The cleaning efficiency of the chemical methods are further improved due to rotating the drill pipe, that causes the turbulence of the fluid in well and improve its debris carrying capacity. Rotation of chemical solution also leads to better cleaning in deviated and horizontal wells by lifting the debris from the underside of the drill pipes. The wellbore cleaning action of chemical fluids is also improved due to turbulent flow achieved due to high speed circulation of fluids.

Nanoemulsions are used as a chemical wellbore cleaning method, where nanoemulsion based spacer fluid is circulated through the drill pipe or casing for removal of existing drilling fluid in the well bore, followed by replacement by pumping of cement slurry. A nanoemulsion based spacer fluid shows better cleanup capability of oil based drilling fluid in comparison to an ordinary emulsion [7]. The improved wellbore cleaning by nanoemulsion is caused mainly due to its smaller droplet size, low IFT, and superior wettability alteration behavior. Thus, filter cake formed by OBM on the formation surface and oil film formed on casing surface is easily removed by the nanoemulsion based spacer fluids [7]. The drilling fluid disperses into the spacer fluid by emulsification, causes efficient drilling fluid removal and wellbore cleaning. Due to the droplet size of 50–500 nm in nanoemulsion based spacer, it can solubilize more oil in comparison to conventional emulsion or surfactant based spacer fluids. The smaller size of nanoemulsions also allows the droplets to enter the pores of filter cake and break the structure of filter cake, causing it to be easily flushed [8].

4.2 Cleaning of Oil Spills

Oil spills is a form of pollution, termed for spillage of crude oil or petroleum products into the environment. Oil spills can happen in marine as well as terrestrial environment. Release of crude oil from offshore platforms into ocean and water bodies while production operations, from tankers into marine ecosystem and leakage of petroleum and refined products from pipelines while transportation, are considered oil spills. Oil spills have disastrous repercussions for environment, economics, and society, thus oil spill incidents usually have immediate and intense media consideration to preventing further spillage and remediation actions [9]. The amount of oil spilled can vary from few hundred tons to thousands of tons, however this can be used to quantify the damage by oil spills. The impact can be over large areas and thousands of meters away from the spill site. The immediate danger after an oil spill is usually a fire breakout which can escalate the damage due to oil spill. Oil spills in terrestrial environment can cause seep onto the ground and contaminate the water bearing zones, leading to pollution of potable water. The vegetation in the surroundings of oil spill is also severely affected, as the percolation of oil near plants can inhibit the uptake of water though roots. The birds and animals confined to area near oil spills can intoxicate themselves by feeding on plants and insects affected by the oil [10]. Oil can also blind the animals and render them defenseless. Oil also masks the scent of the birds and animals, thus leading to abandonment of young birds and animals. Birds which preen have high chances of ingesting oil from their feathers and evolving kidney and liver damage. Ingestion of oil also causes dehydration and weakens digestion process.

Oil spilled in marine systems causes more damage in comparison to spills on land, as oil can easily spread affecting wider area by floating over water bodies. Containment of oil spills on land is also easier than spills in offshore locations. Land animals can avoid the areas affected by oil spills on land in comparison to plants and animals depended on marine ecosystem. Oil penetrates into the feathers of seabirds and fur of the marine mammals, such as sea otters, thus reducing their ability to insulate their bodies and making them more susceptible to temperature fluctuations and less buoyant in water. Dolphins and whales can inhale spilled oil and harm their lungs. Adult fishes can have reduction in their growth, erosion of fins, weakening of reproduction and immunity, due to exposure to oil. Ingestion of oil by zooplanktons, such as mysids, euphausiids, and copepods, leads to their mortality [11]. Narcosis, drowsiness and disruption of central nervous system are major effects of oil spills [12]. Coral reefs which act as nurseries to fishes, shrimps and other animals are also affected by chronic exposure to spilt oil.

The cleaning of oil spills is difficult and expensive, and method selected for cleanup depends on several factors such as type of oil spilled, extent of spill, location of incident, and available resources at the time of the incident [13]. Several physical, chemical, and thermal methods applied for cleaning of oil spills in water bodies are use of oil booms, skimmers, sorbents, dispersants, and burning of floating oil [14]. Oil booms and skimmers are mechanical methods of cleaning of oil spills, where booms contain the spreading of the spilt oil by acting as a fence, as shown in Fig. 4.1. The placement of oil booms around the spill area is effective when the oil is in single spot, applied within few hours of incident, and in areas with low sea waves and tides. Oil contained by oil booms are removed from the water surface by using skimmers, which are designed to extract the oil from the surface of the water. Sorbents, such as peat moss, straw, hay, can also be used which absorbs or adsorbs the oil in the incident of small spills or remove leftovers of larger accidents. Several dispersal chemicals can also be utilized which causes the natural breakdown of components of oil. However, the toxicity of these dispersants to marine ecosystem, should also be considered. Burning of oil can remove most of the oil spill, however, fumes released by burning can be damaging to environment and marine life. Bioremediation is also

Fig. 4.1 Boom deployed to prevent the oil spreading

an effective but time-consuming method, in which the use of microorganisms can degrade oil components into non-toxic compounds, and can be easily applied in extreme weather conditions.

The chemical or physical method applied for the extraction of oil from the sand contaminated during oil spill vary as per the quantity and type of oil and sand. The contamination can spread to wider area for an oil with lower viscosity, spilt on a porous sand [15]. Difficulty in cleaning of oil from sand occurs due to high inter-facial tension of oil and lower solubility of oil components in water, thus, simple flushing with water is ineffective. Thus, surfactant based remediation is used for removal of hydrophobic oil components, but the adsorption of surfactant on sand, silt and clay surface limits its effectiveness [16]. Microemulsions can also be used for remediation of areas contaminated by heavy oil components. Nanoemulsion formu-lated by vegetable oils is also a sustainable and environment friendly method for cleaning of spills in terrestrial regions [17, 18]. The efficiency of nanoemulsion in washing the sand contaminated with oil is tested by determining the residual oil in the sand after treatment with nanoemulsion [19]. Nanoemulsions can penetrate pores and dissolve the oil trapped there because of their smaller droplet size.

4.3 Enhanced Oil Recovery

Enhanced oil recovery (EOR) is the process of recovery of oil trapped in the pores of the reservoir after primary and secondary recovery processes. The EOR techniques includes the processes which change the reservoir rock and fluid properties, such as interfacial tension between oleic and aqueous reservoir fluids, wettability of reservoir rock surface, and viscosity of reservoir fluids, to improve the oil production from the reservoir. These changes in the reservoir rock and fluid properties are achieved by injection of chemicals, or miscible gases, or by increasing the temperature of the reservoir accomplished as a result of injection of steam or hot fluids in the reservoir. The injection of emulsions and nanoemulsions formed by mixing water and hydrocarbons in presence of surfactants are categorized as chemical EOR technique produces the trapped oil from the pores of the reservoir by similar mechanisms such as IFT reduction, wettability alteration, and viscosity modification. Emulsions and nanoemulsions are also miscible with crude oil, which changes the properties of crude oil and improves its flowability through the pores of the reservoir. However, emulsions have poor stability due to its larger droplets, which leads to faster coalescence of the droplets. Thus, injection of emulsions is not significantly effective in recovering the trapped oil from the pores of the reservoir [20]. Nanoemulsions, which have better stability in comparison to emulsions, recovers oil from the pores of the reservoir without resulting to phase separation in reservoir and solubilizing greater volume of trapped oil. The mechanism of EOR by nanoemulsions has been discussed in detail in the following chapter.

References

1. Walters CC (2006) The origin of petroleum. In: Practical advances in petroleum processing. Springer, New York, pp 79–101. https://doi.org/10.1007/978-0-387-25789-1_2
2. Bret-Rouzaut N, Favennec JP (2011) Oil and gas exploration and production: reserves, costs, contracts. Editions Technip
3. Gary JH, Handwerk GE, Kaiser MJ (2007) Petroleum refining: technology and economics, 5th edn. CRC Press
4. Plank J, Tiemeyer C, Buelichen D (2014) A study of cement/mudcake/formation interfaces and their impact on the sealing quality of oilwell cement. Society of petroleum engineering—IADC/SPE Asia Pacific drilling technology conference 2014 driving sustainable growth through technology innovation, pp 1–8. https://doi.org/10.2118/170452-ms
5. Lichinga KN, Luanda A, Sahini MG (2022) A novel alkali-surfactant for optimization of filtercake removal in oil–gas well. J Pet Explor Prod Technol. https://doi.org/10.1007/s13202-021-01438-1
6. Choodesh A, Rutland MG, Grant CWG (2018) Effective approach to wellbore clean out operation: a case study from Zawtika phase 1B. In: Proceedings of IADC/SPE Asia Pacific drilling technology conference APDT, 2018 Aug, pp 1–20. https://doi.org/10.2118/190971-ms
7. Meng R, Wang C, Shen Z (2020) Optimization and characterization of highly stable nanoemulsion for effective oil-based drilling fluid removal. SPE J 25:1259–1271. https://doi.org/10.2118/199904-PA
8. Wang C, Meng R, Xiao F, Wang R (2016) Use of nanoemulsion for effective removal of both oil-based drilling fluid and filter cake. J Nat Gas Sci Eng 36:328–338. https://doi.org/10.1016/j.jngse.2016.10.035
9. Broekema W (2016) Crisis-induced learning and issue politicization in the EU: the BRAER. Sea Empress Erika Prestige Oil Spill Disasters Publ Adm 94:381–398. https://doi.org/10.1111/padm.12170
10. Nelson-Smith A (1979) The effect of oil spills on land and water. In: The prevention of oil pollution. Springer, Netherlands, Dordrecht, pp 17–34. https://doi.org/10.1007/978-94-011-7347-6_2
11. Jiang Z, Huang Y, Xu X, Liao Y, Shou L, Liu J, Chen Q, Zeng J (2010) Advance in the toxic effects of petroleum water accommodated fraction on marine plankton. Acta Ecol Sin 30:8–15. https://doi.org/10.1016/j.chnaes.2009.12.002
12. Baschek B, Gade M, van Bernem K-H, Schwichtenberg F (2015) The German operational monitoring system in the north sea: sensors, methods and example data, pp 161–192. https://doi.org/10.1007/698_2015_399
13. Kemp LK, Lochhead RY, Morgan SE, Savin DA (2013) A new emulsification method: with application for cleaning oil spills. CoatingsTech 10:34–40
14. Hoang AT, Pham VV, Nguyen DN (2018) A report of oil spill recovery technologies. Int J Appl Eng Res 13:4915–4928
15. Lee D-H, Cody RD, Kim D-J, Choi S (2002) Effect of soil texture on surfactant-based remediation of hydrophobic organic-contaminated soil. Environ Int 27:681–688. https://doi.org/10.1016/S0160-4120(01)00130-1
16. Joshi MM, Lee S (1996) Optimization of surfactant-aided remediation of industrially contaminated soils. Energy Sour 18:291–301. https://doi.org/10.1080/00908319608908768
17. Oliveira PF, Spinelli LS, Mansur CRE (2012) The application of nanoemulsions with different orange oil concentrations to remediate crude oil-contaminated soil. J Nanosci Nanotechnol 12:4081–4087. https://doi.org/10.1166/jnn.2012.6166
18. Yap CL, Gan S, Ng HK (2010) Application of vegetable oils in the treatment of polycyclic aromatic hydrocarbons-contaminated soils. J Hazard Mater 177:28–41. https://doi.org/10.1016/j.jhazmat.2009.11.078

19. Oliveira PF, Oliveira TM, Spinelli LS, Mansur CRE (2014) Development and evaluation of solbrax-water nanoemulsions for removal of oil from sand. J Nanomater 2014:1–8. https://doi.org/10.1155/2014/723789
20. Kumar N, Mandal A (2018) Surfactant stabilized oil-in-water nanoemulsion: stability, interfacial tension, and rheology study for enhanced oil recovery application. Energy Fuels 32:6452–6466. https://doi.org/10.1021/acs.energyfuels.8b00043

Chapter 5
Physico-Chemical Properties of Nano-emulsions for Application in EOR

Enhanced oil recovery is the way of recovering crude oil from the reservoir, which remains trapped in the pores of the reservoir after primary and secondary recovery stages. During the primary recovery stage, the crude oil is produced by the natural energy of the reservoir. These natural drive energies are provided by the innate reservoir properties such as compressibility of reservoir rock and fluids, composition of reservoir fluids, presence of gas cap or active aquifer, and density difference of reservoir fluids [1]. The expansion drive mechanism is the dominant mechanism when the reservoir pressure is above the bubble point pressure and the production of reservoir fluids through the wells causes the reduction in reservoir pressure and subsequent expansion of reservoir fluids. The average recovery by the expansion drive mechanism is 3% [2, 3]. The reduction of reservoir pressure below the bubble point pressure leads to the release of the dissolved gases from the crude oil. As the released gases have greater compressibility in comparison to the other reservoir fluids and rock, the expansion of gases becomes the dominant drive mechanism for the production of crude oil. This drive mechanism is termed as solution gas drive or depletion drive and have an average oil recovery of 20% [2]. The crude oil production from the reservoir caused by the expansion of reservoir fluids have lower recovery factor and leads to significant reduction in reservoir pressure. When the reservoir is in communication with a gas cap, present at the top of crude oil-bearing zone, or an aquifer, present at bottom or peripheral of crude oil-bearing zone, the pressure differential between the borehole and reservoir is maintained, leading to significantly higher oil recovery and production life of reservoir.

The natural capacity of a reservoir to produce oil gets depleted with production and requires additional substitutes provided externally by using injection wells. The external energy is usually provided in the form of pressure maintenance or water flooding. The pressure maintenance methodology includes gas injection above gas-oil contact (GOC) or water injection below oil-water contact (OWC), that maintains or increases the pressure of reservoir [4]. The injection of water in the oil zone of

N. Saxena et al., *Nano Emulsions in Enhanced Oil Recovery*,
SpringerBriefs in Petroleum Geoscience & Engineering,
https://doi.org/10.1007/978-3-031-06689-4_5

the reservoir is termed as water flooding, where the injected water sweeps the hydro-carbons from the reservoir towards the production well. The oil recovery from the reservoir by the use of external aid is termed as secondary oil recovery. The primary and secondary oil recovery stages do not incorporate any alteration of reservoir rock and fluid properties leading to the trapping of crude oil in the reservoir.

The unrecovered crude oil is trapped in the pores of the reservoir due to various factors such as high capillary forces acting on the oil droplets caused due to the high IFT of the oil-aqueous interface. The trapping of crude oil due to high IFT is more prominent in water-wet reservoirs, whereas in oil-wet reservoirs, the residual oil is present as oil films coated on the rock surfaces. Thus, in an oil-wet reservoir, the oil is stuck in the smaller pores leading to greater residual oil saturation. The oil droplets are also trapped at the "dead ends" of the flow channels, or in microscopic pores of a heterogeneous reservoir [5]. Thus, the reservoir rock and fluid properties such as pore shape and size, reservoir heterogeneities, IFT of the reservoir fluids, and the wetting state of the reservoir rock are the major factors causes the low oil recovery efficiency of 30–40% of original oil in place (OOIP) after primary and secondary recovery stages [6, 7]. Thus, the manipulation and tuning of reservoir conditions, that include rock properties, fluid properties, and reservoir temperature, are applied in the tertiary oil recovery stage to recover the oil trapped in the reservoir and this oil recovery stage is termed as enhanced oil recovery (EOR). The stages and classifications of oil recovery methods have been shown in Fig. 5.1.

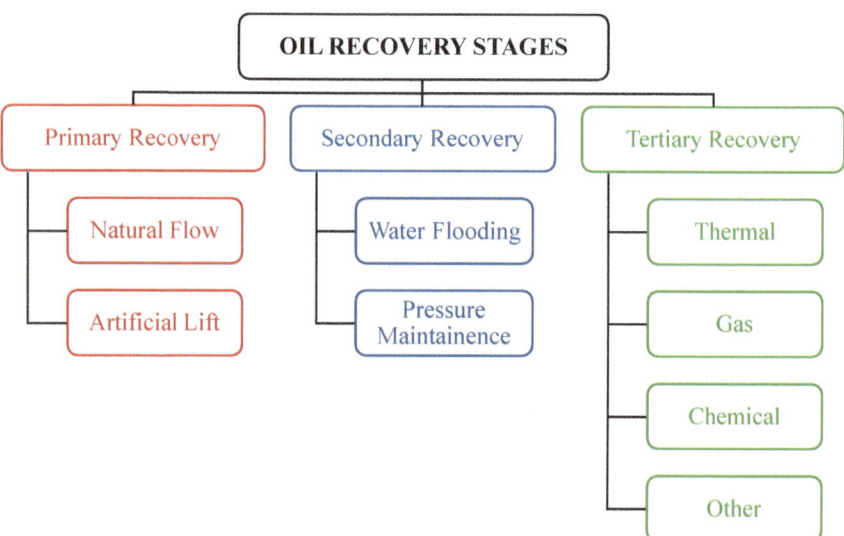

Fig. 5.1 Classification of oil recovery

5.1   Enhanced Oil Recovery (EOR)

A typical EOR process involves injection of EOR solution through the injection well, followed by chase water flooding that pushes the EOR solution towards the production well, leading to the recovery of trapped oil after primary and secondary recovery stages. The general schematic of EOR process has been illustrated in Fig. 5.2.

5.1.1   Principles of EOR

EOR techniques aim to recover the crude oil trapped in the reservoir after the primary and secondary production stages. The unrecovered crude oil is trapped in the pores of the reservoir due to various reservoir rock and fluid properties, such as high IFT between the oleic phase and aqueous phase present in the reservoir, oil-wetting state of reservoir rock surface, and high viscosity of crude oil. These properties lead to lower microscopic and macroscopic efficiencies of the crude oil production by the natural and external energies provided to the reservoir. Thus, EOR application focuses on the oil production by principles such as IFT reduction, wettability alteration, and improvement of mobility ratio by injection fluid. The reduction of IFT between

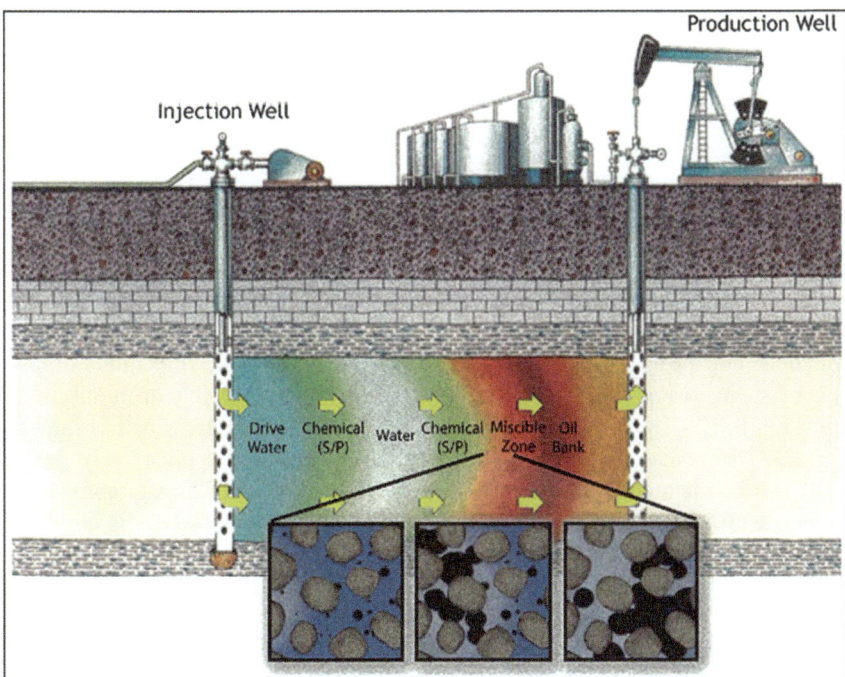

Fig. 5.2 Schematic illustration of enhanced oil recovery [8]

the oleic phase and aqueous phase present in the reservoir leads to the reduction of capillary forces causing the mobilization of trapped crude oil droplets from the narrow pores of the reservoir. The reduction of capillary forces that traps the crude oil in the reservoir pores is also caused by the wettability alteration of reservoir rock surface from oil-wetting state to preferentially water-wetting state. The reduction of capillary force by IFT reduction and wettability alteration leads to mobilization of crude oil from the reservoir to the production well. Thus, IFT reduction and wettability alteration cause the increase in the microscopic sweep efficiency of the applied EOR technique. The mobilization of crude oil from the microscopic pores of the reservoir also leads to the formation of oil bank, which is pushed towards the production well by the injection of chase fluid after EOR application. However, the efficient mobilization of oil bank towards the production well is also a factor of the macroscopic sweep of the EOR techniques. The macroscopic efficiency of the EOR method is improved by controlling the mobility ratio of displacing fluid and displaced fluid. A lower mobility ratio is preferred for efficient volumetric sweep of the crude oil. This lower mobility ratio of displacing fluid to crude oil is achieved by either reduction of crude oil viscosity of crude oil or increase in the viscosity of displacing fluid. Lower mobility ratio improves macroscopic efficiency of EOR by preventing early breakthrough of injected fluid and improving the effective swept volume of reservoir. The macroscopic efficiency, also known as volumetric sweep efficiency (E_{vo}), and microscopic efficiency, also known as displacement efficiency (E_d), is related to the overall efficiency (E_o) of EOR method as:

$$E_o = E_d \times E_{vo} = E_d \times E_a \times E_v \qquad (5.1)$$

where, E_a and E_v are the areal and vertical sweep efficiencies.

5.1.2 Classifications of EOR

The various oil recovery mechanisms of EOR are achieved by different techniques, which include the use of chemicals in injected fluid, increasing temperature of reservoir by steam or hot water injection, and injecting gases miscible or immiscible to crude oil [9, 10]. These different techniques used for EOR are classified as shown in Fig. 5.3. The increase in reservoir temperature is achieved by application of thermal EOR, where methodology of injection of steam or hot water, in-situ combustion, cyclic steam injection, and steam-assisted gravity drainage (SAGD) is employed. The increase in temperature achieved by these techniques causes the reduction of crude oil viscosity leading to the enhancement of the mobility of the trapped crude oil droplets. Thus, thermal EOR is mostly applicable to reservoirs with heavy oil [11]. The injection of gases such as natural gas, CO_2 and N_2 are used in gas EOR process, which leads to oil recovery. These gases are either miscible or immiscible with the reservoir fluids. The immiscible gases push the oil to production well by

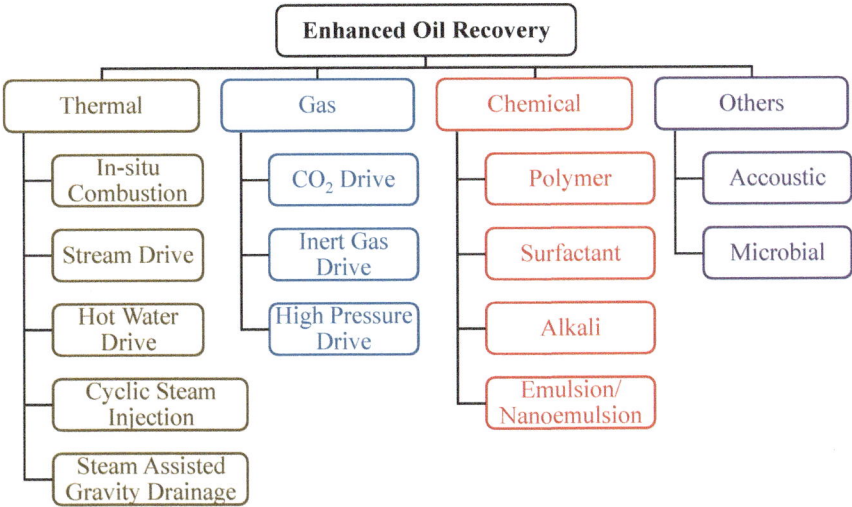

Fig. 5.3 Conventional EOR classification

expansion in reservoir, whereas the miscible gases reduce the viscosity of crude oil after getting dissolved in the trapped crude oil, thus leading to improvement in oil flowability. Thus, application of gas EOR is widely implemented in reservoirs with light to moderate oil [12]. Chemical EOR includes injection of chemicals such as polymer, surfactant, alkali, and emulsion into the reservoir to improve the oil recovery. Polymer injection leads to the increase viscosity of injected fluid causing the viscous forces acting on the trapped crude oil. The injection of surfactant and alkali causes the reduction of IFT between the displaced and displacing fluids, leading to decrease of capillary forces that trap the crude oil in the pores of the reservoir. The injection of emulsions or nanoemulsions into the oil reservoirs enhances the oil production due to their favorable rheological and thermodynamic properties [13]. Thus, combination of different chemicals is used in chemical EOR for recovery of oil from the reservoir. Other EOR methodology, that includes using microbial and acoustic application, are in experimental stage and lacks the commercial large-scale implementation in reservoirs [14].

5.2 Nanoemulsions in EOR

Nanoemulsions are dispersions of oleic and aqueous phase, which are stabilized by surfactant molecules adsorbed at the interface of oleic and aqueous phases. The dispersions of nanoemulsions are thermodynamically stable, isotropic, and transparent or translucent, with droplet size of 500 nm or less [15, 16]. Nanoemulsions can spontaneously solubilize more amount of crude oil and reduce the IFT between oil and water to ultralow values [17]. The physico-chemical properties of nanoemul-sions imply that it can used to recover the trapped crude oil from the pores of the reservoir [18].

5.2.1 Interfacial Tension (IFT)

IFT is an important parameter that affects the oil displacement through the pores of the reservoir. Higher IFT between oil and displacing fluid signifies a rigid interface between oil and aqueous phase, which restrict the large oil globules to pass through the narrow pores of the reservoir. This is also evident from the higher capillary pressure as a result of high IFT, which needs to be overcome by the viscous forces to allow for the flow of oil in the reservoir [19]. IFT of the oil-water interface during conventional water flooding is usually 10–30 mN/m [20]. The shape of crude oil for high IFT condition has been represented in Fig. 5.4a. Thus, the high IFT between the fluids in the reservoir leads to lower microscopic efficiency of primary and secondary stages of oil recovery. During the displacement of crude oil droplets by the injected aqueous slug, lowering of IFT between the oleic and aqueous phases allow the oil-water interface to easily deform, ensuring the passage of the trapped oil through the narrow pores of the reservoir due to the viscous forces of the injected slug. The shape of crude oil droplet for low IFT condition has been represented in Fig. 5.4b.

When a nanoemulsion is selected for EOR, it should have low IFT with the crude oil as well as the water phase. Lower IFT between the crude oil and nanoemul-sion ensures that the nanoemulsion, acting as the displacing fluid, can effectively displace the trapped crude oil. Similarly, low IFT between the nanoemulsion slug and chase water guarantees that nanoemulsions is also displaced through the pores of the reservoir by the chase water. IFT between two immiscible phases can be measured by pendant drop method or spinning drop method. Spinning drop method can be used to determine the IFT between the phases in the range of 10^{-3} mN/m, however, this method involves spinning of a droplet of lighter phase (nanoemulsion) surrounded by heavier phase (water). As the nanoemulsions used as EOR agents are mostly oil in water emulsions, thus measurement of IFT between nanoemulsion and water is practically impossible by spinning drop technique. The pendent drop

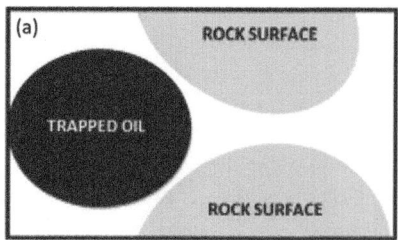
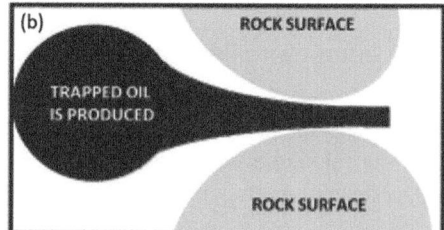

Fig. 5.4 Schematic of effect of **a** high IFT and **b** low IFT on trapped crude oil [21]

method for determination of IFT also suffers similar limitation while measurement of IFT between nanoemulsion and water. However, IFT between hydrocarbon and nanoemulsion can be determined by using pendent drop method, where droplet of nanoemulsion is suspended at the tip of a needle surrounded by the bulk hydrocarbon phase. IFT in the range of 1–5 mN/m have been determined for nanoemulsion and pure hydrocarbon phase [22]. IFT between crude oil and nanoemulsion is in the range 10^{-3} mN/m or lower, thus it easily solubilizes significant amount of trapped oil, leading to its production due to nanoemulsion flooding [18].

5.2.2 Miscibility

Miscibility of crude oil with injected emulsion or nanoemulsion occurs in the pores of the reservoir due to low IFT between the nanoemulsion and crude oil. Miscibility of the trapped crude oil and nanoemulsion can improve the microscopic oil recovery efficiency of the injected EOR slug. Miscibility of nanoemulsion with crude oil also ensures that the separation of phases does not occurs in the pores of the reservoir, and the injected nanoemulsion does not blocks the pores of the reservoir. The single phase formed by the miscibility of the crude oil and nanoemulsion will have lower viscosity due to increase in its saturate content [23]. The miscibility behaviour should also be similar for different proportions of crude oil and nanoemulsion, so the effect of varying residual oil saturation is minimal and lower value of residual oil is achieved after nanoemulsion flooding [24].

5.2.3 Wettability Alteration

The wetting state of a reservoir rock has a significant effect on the oil recovery of crude oil from the reservoir. The wetting state has been defined on the basis of the contact angle of the aqueous phase on the rock surface in a three-phase system consisting of rock surface, aqueous phase and oleic phase. Contact angle between 0° and 30° corresponds to a strongly water-wet surface, whereas, the contact angle between 30° and 75° signifies a moderately water-wet surface. A neutral wet surface forms contact angle between 75° and 105° and moderately oil-wet surface forms contact angle between 105° and 150°. Contact angle between 150° and 180° is formed on a strongly oil-wet surface [25].

The wetting state of a reservoirs were initially water-wet as the pores of the reservoir rock were initially occupied with water/brine. The crude oil that migrated into the pores of the reservoir from their primary sources interacted with the reservoir rock surface and led to the change in the wetting state of the reservoir [26]. Thus, the reservoir rocks were initially water-wet in nature and the prolonged interaction of crude oil with reservoir rock caused the adsorption of polar components of crude

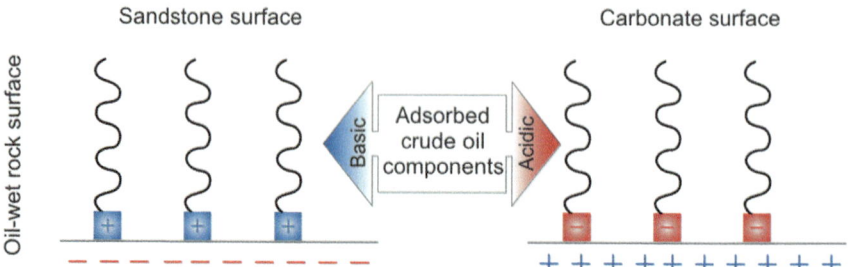

Fig. 5.5 Schematic of adsorbed crude oil components causing oil wetting of rock surfaces [30]

oil on the rock surface, leading to change in wetting state of reservoir rock to oil-wetting state [27]. However, the interaction of the crude oil with the surface of the reservoir rock is prevented by the film of water present between the rock/water and oil/water interfaces. The strength of the water film is dependent on mineralogy of reservoir rock, constituents of crude oil and brine in the reservoir [28]. A strong water film maintains the water-wetting state of the reservoir, whereas a weak water film is easily ruptured, leading to contact of crude oil and rock surface. The existence of attractive forces such as electrostatic contact, van der Waals interactions, and hydrogen bonding between crude oil and the rock surface causes the brine layer to rupture [29]. Thus, these forces cause crude oil components to adsorb on the rock surface and the reservoir rock to become oil-wetting. Figure 5.5 shows the schematic representing the adsorbed crude oil components leading to oil wetting of the reservoir rock surface.

The rock mineralogy of a reservoir rock also plays an important role in the alteration of rock wettability from water-wet to oil-wet. The minerals present in the reservoir rock leads to the charge at their surface. The surface of sandstone reservoir rock is negatively charged due to presence of minerals such as quartz [31]. The surface of carbonate reservoir rock is positively charged due to the presence of minerals such as calcite and dolomite [32]. These charged surfaces attract the polar components of the crude oil. The negatively charged sandstone rock surface attracts the positively charged basic components of the crude oil. Similarly, attractive forces exist between the positively charged carbonate rock surface and the negatively charged acidic components of the crude oil [33, 34]. Thus, the adhesion of the crude oil components on the rock surface causes the oil-wetting of the reservoir rocks.

A reservoir with oil-wet rock surface has lower oil recovery as the crude oil occupies the smaller pores in an oil-wet reservoir, leading to the lower mobility of oil under the influence of pressure differential in the reservoir [35]. Thus, alteration of wetting state from oil-wet to water-wet is desired for the increase in production of crude oil. The injection of nanoemulsion leads to the formation of ion-par between the surfactant molecules present in the nanoemulsion and adsorbed crude oil components on the surface of the reservoir rock [36]. The crude oil desorbed from the rock surface is also solubilized into oleic phase of nanoemulsion [24]. This detached crude oil from the surface is released in to the injected nanoemulsion, leading to enhanced

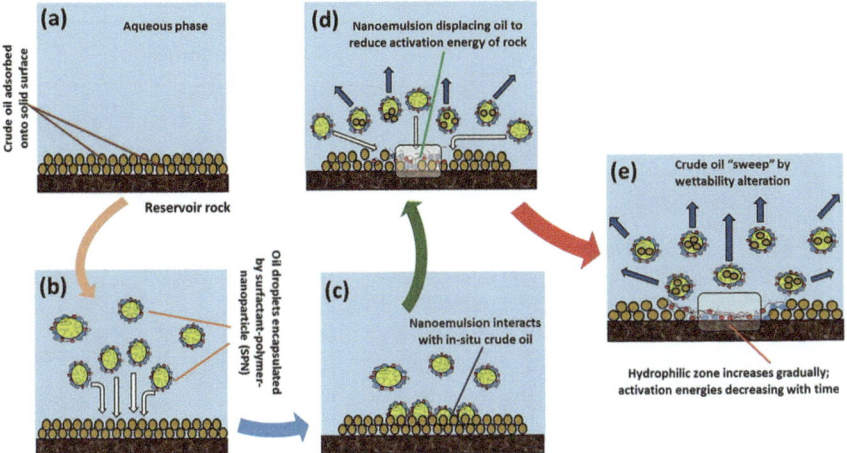

Fig. 5.6 Schematic of wettability alteration and oil production by nanoemulsion [22]

oil production [37]. The production of crude oil by wettability modification of rock surface by nanoemulsions has been depicted in Fig. 5.6. Nanoemulsions initially interacts with the adsorbed components of crude oil on the surface of the reservoir rock and surfactant aids in the detachment of crude oil components into the aqueous phase.

5.2.4 Mobility Control

Mobility control by the injected fluid is an important parameter that regulates the macroscopic sweep efficiency of the EOR fluid. High viscosity of injected fluid leads to reduction of its mobility and improves the mobility ratio of displacing fluid and displace fluid [38]. The waterflood done in secondary recovery stage have lower volumetric sweep efficiency due to the lower viscosity and higher mobility of water that causes various problems such as fingering and early breakthrough of injected fluid from the production well through the high permeable streaks in the heterogeneous reservoirs [39]. The displacing EOR fluid having lower mobility ratio with the displaced fluid forms a stable flood front as viscous injected fluid moves at the lower flow rate in high permeable zones. Thus, the oil trapped in the lower permeable zones is also displaced by the injected fluid, thus preventing fingering and early breakthrough of the injected EOR slug.

Nanoemulsions have favorable viscosity and better rheological properties in comparison to microemulsions. Nanoemulsions can have varied rheological behaviour, such as Newtonian, shear thinning, or shear thickening behaviour. The different methodologies of formulation of nanoemulsions, which gives control over

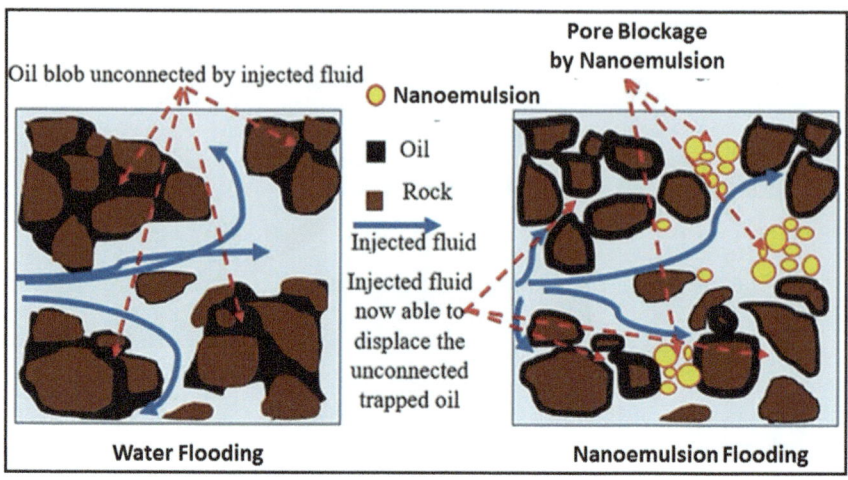

Fig. 5.7 Schematic of plugging of high permeable zones by nanoemulsion flooding [42]

the droplet size, volume fraction of dispersed and continuous phase, and use of additives such as viscosifiers, allows for formulation of nanoemulsions with desirable viscosity and rheological properties [40]. The rheological behaviour of nanoemulsions is broadly defined by Power Law model and Herschel-Bulkley Model. The shear thinning behaviour of nanoemulsions is due to the reduction of droplet-droplet interaction caused due to decrease in the Brownian motion and realignment of droplets leading to lower resistance to movement at higher shear rate. The rheological properties of nanoemulsions alter fluid flow within reservoir pores, resulting in increased migration, swept volume, and displacement efficiency in porous media. The rheological behavior of nanoemulsions makes them a suitable chemical slug for mobility control. The capillary forces that confine crude oil at its site are influenced by flooding nanoemulsions into the reservoir [41]. During the injection of nanoemulsion, the fluid flows under high shear condition and the nanoemulsion structure gets disturbed. When the nanoemulsion enters the reservoir pores, it flows under low shear condition and the nanoemulsion structure is reformed. Thus, nanoemulsion does not suffers with shear degradation, which is a common problem for polymer based EOR fluids. Nanoemulsion recover the residual oil due to its ability to sidetrack the fluid flow from high permeable regions to low permeable regions. This phenomenon is also depicted in the schematic shown in Fig. 5.7.

References

1. Speight JG (2016) General methods of oil recovery. In: Introduction to enhancing recover. In: Methods heavy oil tar sands. Elsevier, pp 253–322. https://doi.org/10.1016/B978-0-12-849906-1.00006-0

2. Satter A, Iqbal GM (2016) Primary recovery mechanisms and recovery efficiencies. In: Reservoir engineering. Elsevier, pp 185–193. https://doi.org/10.1016/B978-0-12-800219-3.00011-5

3. Iglauer S, Wu Y, Shuler P, Tang Y, Goddard WA (2010) New surfactant classes for enhanced oil recovery and their tertiary oil recovery potential. J Pet Sci Eng 71:23–29. https://doi.org/10.1016/j.petrol.2009.12.009

4. Vishnyakov V, Suleimanov B, Salmanov A, Zeynalov E (2020) Oil recovery stages and methods. In: Primer on enhanced oil recovering. Elsevier, pp 53–63. https://doi.org/10.1016/B978-0-12-817632-0.00007-4

5. Sheng JJ (2011) Modern chemical enhanced oil recovery. Elsevier. https://doi.org/10.1016/C2009-0-20241-8

6. Fink J (2015) Petroleum engineer's guide to oil field chemicals and fluids. Elsevier. https://doi.org/10.1016/C2015-0-00518-4

7. Austad T, Hodne H, Strand S, Veggeland K (1996) Chemical flooding of oil reservoirs 5. The multiphase behavior of oil/brine/surfactant systems in relation to changes in pressure, temperature, and oil composition. Colloid Surf A Physicochem Eng Asp 108:253–262. https://doi.org/10.1016/0927-7757(95)03405-6

8. Saxena N, Mandal A (2022) Natural surfactants. Springer International Publishing, Cham. https://doi.org/10.1007/978-3-030-78548-2

9. Srivastava RK, Huang SS, Dong M (1999) Comparative effectiveness of CO_2 produced gas, and flue gas for enhanced heavy-oil recovery. SPE Reserv Eval Eng 2:238–247. https://doi.org/10.2118/56857-PA

10. Elgaghah S, Zekri AY, Almehaideb RA, Shedid SA (2007) Laboratory investigation of influences of initial oil saturation and oil viscosity on oil recovery by CO_2 miscible flooding. In: European conference exhibition. Society of Petroleum Engineers. https://doi.org/10.2118/106958-MS

11. Taber JJ, Martin FD, Seright RS (1997) EOR screening criteria revisited—part 1: introduction to screening criteria and enhanced recovery field projects. SPE Reserv Eng 12:189–198. https://doi.org/10.2118/35385-PA

12. Mahdavi E, Zebarjad FS (2018) Screening criteria of enhanced oil recovery methods. In: Fundamentals of enhanced oil gas recovering from conventional and unconventional reservoirs. Elsevier, pp 41–59. https://doi.org/10.1016/B978-0-12-813027-8.00002-3

13. Mohyaldinn ME, Hassan AM, Ayoub MA (2019) Application of emulsions and microemulsions in enhanced oil recovery and well stimulation. In: Microemulsion—a chemical nanoreactor, IntechOpen. https://doi.org/10.5772/intechopen.84538

14. Yernazarova A, Kayirmanova G, Baubekova A, Zhubanova A (2016) Microbial enhanced oil recovery. In: Chemical enhanced oil recovery—a practical overview, InTech, pp 295–304. https://doi.org/10.5772/64805

15. Del Gaudio L, Bortolo R, Lockhart TP (2007) Nanoemulsions: a new vehicle for chemical additive delivery. In: Proceedings of SPE international symposium oilfield chemistry, pp 212–220. https://doi.org/10.2523/106016-ms

16. Mandal A, Bera A, Ojha K, Kumar T (2012) Characterization of surfactant stabilized nanoemulsion and its use in enhanced oil recovery. In: SPE international oilfield nanotechnology conference on exhibition. Society of Petroleum Engineers, pp 537–542. https://doi.org/10.2118/155406-MS

17. See CH, Saphanuchart W, Nadarajan S, Lim CN (2018) Nanoemulsion for non-aqueous mud removal in wellbore. Soc Pet Eng SPE/DGS Saudi Arab Sect Tech Symp Exhib 2011:700–709. https://doi.org/10.2118/149088-ms

18. Hashimah Alias N, Aimi Ghazali N, Amran Tengku Mohd T, Adieb Idris S, Yahya E, Mohd Yusof N (2015) Nanoemulsion applications in enhanced oil recovery and wellbore cleaning: an overview. In: Applied mechanics and materials, vol 754–755, pp 1161–1168. https://doi.org/10.4028/www.scientific.net/amm.754-755.1161

19. Kumar A, Mandal A (2018) Characterization of rock-fluid and fluid-fluid interactions in presence of a family of synthesized zwitterionic surfactants for application in enhanced oil recovery. Colloid Surf A Physicochem Eng Asp 549:1–12. https://doi.org/10.1016/j.colsurfa.2018.04.001

20. Ahmed S, Elraies KA (2018) Microemulsion in enhanced oil recovery. In: Science technology behind nanoemulsions. InTech, p 13. https://doi.org/10.5772/intechopen.75778

21. Kumar A, Mandal A (2017) Synthesis and physiochemical characterization of zwitterionic surfactant for application in enhanced oil recovery. J Mol Liq 243:61–71. https://doi.org/10.1016/j.molliq.2017.08.032

22. Pal N, Mandal A (2020) Oil recovery mechanisms of Pickering nanoemulsions stabilized by surfactant-polymer-nanoparticle assemblies: a versatile surface energies' approach. Fuel 276:118138. https://doi.org/10.1016/j.fuel.2020.118138

23. Alomair OA, Almusallam AS (2013) Heavy crude oil viscosity reduction and the impact of asphaltene precipitation. Energy Fuels 27:7267–7276. https://doi.org/10.1021/ef4015636

24. Kumar A, Saw RK, Mandal A (2019) RSM optimization of oil-in-water microemulsion stabilized by synthesized zwitterionic surfactant and its properties evaluation for application in enhanced oil recovery. Chem Eng Res Des 147:399–411. https://doi.org/10.1016/j.cherd.2019.05.034

25. Fanchi JR (2018) Rock–fluid interaction. In: Principles of applied reservoir simulations. Elsevier, pp 81–99. https://doi.org/10.1016/B978-0-12-815563-9.00005-7

26. Rahbar M, Roosta A, Ayatollahi S, Ghatee MH (2012) Prediction of three-dimensional (3-D) adhesion maps, using the stability of the thin wetting film during the wettability alteration process. Energy Fuels 26:2182–2190. https://doi.org/10.1021/ef202017a

27. Saxena N, Kumar A, Mandal A (2019) Adsorption analysis of natural anionic surfactant for enhanced oil recovery: the role of mineralogy, salinity, alkalinity and nanoparticles. J Pet Sci Eng 173:1264–1283. https://doi.org/10.1016/j.petrol.2018.11.002

28. Sohal MA, Thyne G, Søgaard EG (2016) Review of recovery mechanisms of ionically modified waterflood in carbonate reservoirs. Energy Fuels 30:1904–1914. https://doi.org/10.1021/acs.energyfuels.5b02749

29. Hirasaki GJ (1991) Wettability: fundamentals and surface forces. SPE Form Eval 6:217–226. https://doi.org/10.2118/17367-PA

30. Kumar A, Mandal A (2019) Critical investigation of zwitterionic surfactant for enhanced oil recovery from both sandstone and carbonate reservoirs: adsorption, wettability alteration and imbibition studies. Chem Eng Sci 209:115222. https://doi.org/10.1016/j.ces.2019.115222

31. Santha N, Cubillas P, Saw A, Brooksbank H, Greenwell HC (2017) Chemical force microscopy study on the interactions of COOH functional groups with kaolinite surfaces: Implications for enhanced oil recovery. Minerals 7:250. https://doi.org/10.3390/min7120250

32. Derkani MH, Fletcher AJ, Fedorov M, Abdallah W, Sauerer B, Anderson J, Zhang ZJ (2019) Mechanisms of surface charge modification of carbonates in aqueous electrolyte solutions. Colloid Interf 3:62. https://doi.org/10.3390/colloids3040062

33. Mohammed MA, Babadagli T (2016) Experimental investigation of wettability alteration in oil-wet reservoirs containing heavy oil. SPE Reserv Eval Eng 19:633–644. https://doi.org/10.2118/170034-PA

34. Doust AR, Puntervold T, Austad T (2011) Chemical verification of the EOR mechanism by using low saline/smart water in sandstone. Energy Fuels 25:2151–2162. https://doi.org/10.1021/ef200215y

35. Kuliničᵛ V (2015) The influence of wettability on oil recovery. AGH Drilling Oil Gas 32:493. https://doi.org/10.7494/drill.2015.32.3.493

36. Jarrahian K, Vafie-Sefti M, Ayatollahi S, Moghadam F, Moghadam AM (2010) Study of wettability alteration mechanisms by surfactants. In: International symposium society core of analysis, pp 1–6

37. Kopanichuk IV, Vanin AA, Brodskaya EN (2017) Disjoining pressure and structure of a fluid confined between nanoscale surfaces. Colloid Surf A Physicochem Eng Asp 527:42–48. https://doi.org/10.1016/j.colsurfa.2017.04.072
38. Shaker Shiran B, Skauge A (2013) Enhanced oil recovery (EOR) by combined low salinity water/polymer flooding. Energy Fuels 27:1223–1235. https://doi.org/10.1021/ef301538e
39. Kumar M, Hoang VT, Satik C, Rojas DH (2008) High-mobility-ratio waterflood performance prediction: challenges and new insights. SPE Reserv Eval Eng 11:186–196. https://doi.org/10.2118/97671-PA
40. Fryd MM, Mason TG (2012) Advanced nanoemulsions. Annu Rev Phys Chem 63:493–518. https://doi.org/10.1146/annurev-physchem-032210-103436
41. Farid Ibrahim A, Nasr-El-Din H (2018) Stability improvement of CO_2 foam for enhanced oil recovery applications using nanoparticles and viscoelastic surfactants. In: SPE Trinidad Tobago section energy resources conference, SPE. https://doi.org/10.2118/191251-MS
42. Kumar N, Verma A, Mandal A (2021) Formation, characteristics and oil industry applications of nanoemulsions: a review. J Pet Sci Eng 206:109042. https://doi.org/10.1016/j.petrol.2021.109042

Chapter 6
Applications of Nanoemulsions in EOR

The physico-chemical properties of nanoemulsions such as ultralow IFT with crude oil and chase water, miscibility with crude oil, modification of wetting state of reservoir rock surface to water-wetting, and higher mobility control of the injected nanoemulsions makes it a prospective EOR fluid. However, proper screening of the formulated nanoemulsion, its composition and evaluation of formulated nanoemulsion is required for achieving the optimum oil recovery efficiency by nanoemulsion flooding in the reservoir.

6.1 Screening and Potential Evaluation of Nanoemulsions

Screening of applicability of an EOR method is an important step while choosing the best recovery method. With increasing research in the field of EOR, the choice of fluid has also become abundant, with compatibility to application in varied categories and operational conditions of the reservoir rock and fluids. The selection of most profitable EOR fluid would become extremely wearing, if one had to test the potential of every available EOR technique. Thankfully, over time, the screening criteria for selecting EOR have developed, assisting petroleum engineers in making the best decision possible. The screening criteria is based on several reservoir rock and fluid properties, such as, oil density, oil viscosity, composition, saturation, formation type, reservoir thickness, permeability, depth, and temperature [1]. The EOR methods selected based on the screening criteria have higher chance of success, leading to a profitable project, which is a major factor due to varying crude oil prices. The screening criteria applied for much most EOR techniques have evolved based on its applications in reservoirs with vide range of properties, however, the application of nanoemulsions as an EOR agent is new and limited information is available regarding the success of nanoemulsion flooding. Most research based on feasibility of nanoemulsion as EOR agent is still in laboratory stage. Kumar et al. [2] have

© The Author(s), under exclusive license to Springer Nature Switzerland AG 2022   43
N. Saxena et al., *Nano Emulsions in Enhanced Oil Recovery*,
SpringerBriefs in Petroleum Geoscience & Engineering,
https://doi.org/10.1007/978-3-031-06689-4_6

Table 6.1 Screening criteria for application of nanoemulsion in EOR [2]	Oil properties	Gravity (°API)	22.3–35
		Viscosity (cp)	< 35
		Composition	Light/medium oil
	Reservoir characteristics	Type of formation	Sandstone preferred
		Porosity (%)	10–35
		Oil saturation (%)	28–35
		Permeability (mD)	10–350
		Net thickness	Not critical
		Depth (ft)	625–5300
		Temperature (°F)	< 200

provided a screening criterion for evaluation of nanoemulsion flooding in reservoir based on literature, as shown in Table 6.1.

6.2 Oil Mobilization and Recovery Process

The mechanism of recovery of trapped crude oil by the injection of nanoemulsion is similar to the recovery mechanism by emulsion or microemulsions. However, the oil recovery efficiency of nanoemulsion flooding is superior due to its better physico-chemical properties as discussed in the previous chapter. The injection of nanoemulsions in reservoir leads to oil recovery at microscopic as well as macroscopic level of reservoir. The macroscopic recovery efficiency of the injected EOR fluid is given in terms of volumetric sweep efficiency, which relates to the areal and vertical coverage of reservoir by the injected slug. The value closer to one for mobility ratio of displacing to displaced fluid is suggested to have a higher volumetric sweep efficiency. The higher viscosity of nanoemulsion leads to the desired value of mobility ratio and contacts with high permeable as well as low permeable zone of the reservoir. Thus, nanoemulsion flooding ensures the higher value of macroscopic efficiency of the flooding process. At the microscopic scale, the injected nanoemulsion slug recovers the trapped oil from the pores of the reservoir due to its lower IFT between the slug and crude oil. Thus, the nanoemulsion is able to recover the oil, as the droplet of oil can be easily deformed and passed through the pores of the reservoir. For nanoemulsion slug, which have ultralow IFT with the crude oil, the miscibility of oil and injected nanoemulsion can also occur forming a single phase in the pores. The miscible phase formed can easily pass through the pores due to diminished interface and reduced capillary forces keeping the oil droplet trapped in the pores of the reservoir. The oil droplets can also be trapped in pores due to oil wetting nature of the reservoir rock, where the oil droplet is adsorbed on the rock surface and resist flow during primary and secondary recovery processes. The oil droplets adsorbed on the rock surface also gets mobilized due to miscibility with the nano sized oil droplets in the injected nanoemulsion. The surfactant stabilizing the nanoemulsion also gets adsorbed in the rock surface and prevents re-adsorption of the crude oil components. Thus, wettability alteration of the reservoir rock towards water wetting state also aids in the increase in the microscopic sweep efficiency of the nanoemulsion flooding process. The nanosized oil droplets in the nanoemulsion also has the advantage to emulsion and microemulsion flooding processes, as these oil droplets can easily enter the pores throat of nanoscales and recover the oil trapped in nanosized pores of the reservoir, thus reducing the residual oil saturation after nanoemulsion flooding to very low value. The low IFT between the nanoemulsion and aqueous phase also aids to increase in the sweep efficiency of the nanoemulsion, as the single miscible phase formed by the interaction of nanoemulsion slug and trapped crude oil can easily flow through the pores due to viscous forces exerted due to injection of chase fluid. Thus, the overall oil recovery efficiency of nanoemulsion flooding is higher due to higher macroscopic as well as microscopic sweep efficiencies.

6.3 Challenges in Application of Nanoemulsions in EOR

The potential of nanoemulsion as an efficient EOR fluid is implied by its favourable physico-chemical properties. However, the field applicability does face the economic concerns regarding the formulation of nanoemulsions and other large-scale facilities required near the well site. Large amount of water and oleic phase selected for nanoemulsion and the chemicals such as surfactants, viscosifiers, and solvents used for enhancing the properties of nanoemulsions are required. The emulsification process also requires large amount of energy so desired nanoemulsion can be formulated. So, energy-efficient formulation process is required for economic feasibility of application of nanoemulsion as an EOR fluid. The injection of formulated nanoemulsion into the reservoir through injection wells also requires significant investment, that depends on number of injector wells, depth of reservoir, surface facilities for injection of nanoemulsions, as well as sub-surface tools for monitoring the nanoemulsions near the perforations.

Apart from economic viability, the most significant obstacle to using nanoemulsion as an efficient EOR fluid would be its stability in reservoir pores. Once the nanoemulsion enters in the pores of the reservoir, its destabilization can occur due to interaction with reservoir fluids. The change in salinity, temperature and crude oil components can affect the nano size of the dispersed phase and lead to phase separation in the reservoir pores. Thus, thorough understanding of interaction of formulated nanoemulsion with reservoir fluid under harsh reservoir conditions is required for success of the nanoemulsion flooding project. The oil recovery process by nanoemulsion is also widely observed at laboratory scale experiments, thus several other mechanisms occurring in the reservoir, such as dynamic stability of nanoemulsion during long duration of continuous flow through porous media from injection well to production well, interaction of nanoemulsion with variation in the reservoir rock mineralogy, cannot be predicted with high certainty based on lab scale results. This lag in field scale pilot testing of nanoemulsion flooding is necessary for utilization of laboratory results.

References

1. Taber JJ, Martin FD, Seright RS (1997) EOR screening criteria revisited—part 1: introduction to screening criteria and enhanced recovery field projects. SPE Reserv Eng 12:189–198. https://doi.org/10.2118/35385-PA
2. Kumar N, Verma A, Mandal A (2021) Formation, characteristics and oil industry applications of nanoemulsions: a review. J Pet Sci Eng 206:109042. https://doi.org/10.1016/j.petrol.2021.109042

Conclusions

The ability to understand and regulate material qualities at the atomic level is what propels technological advancement. While nanoscience will impact our future, the utilisation of micrometric items like droplets in emulsions has already changed our daily lives. Emulsions are used in a wide range of processes and products, therefore mastering the phenomena that regulate their behaviour is essential. Emulsion science has progressed to the point that more daring applications may now be imagined. Emulsions have found their applications in almost every area; the promising ones are petroleum industry, cosmetic industry, food and preservative industry, detergents and adhesives, inks, coatings and paint, and pharmaceutical industries.

Nano and micro-technologies, in particular, are increasingly benefiting from this foundation, despite the relatively fresh input of emulsions, and this journey is still in its early stages. Indeed, emulsions as intermediate materials provide unique benefits for developing multi-function colloidal particles due to the great variety given by fragmentation methods. Nanoemulsions exhibit a number of distinct characteristics, that includes tiny droplet size, high stability, transparency, and adjustable rheology. Nanoemulsions are an appealing possibility for use in the food, cosmetics, pharmaceutical, and drug delivery sectors because of their distinct characteristics. As substrates, reservoirs, reactors, templates, or simply intermediaries, emulsions seem to have successfully met the nano and micro-technologies adventure.

Nanoemulsions have several applications in petroleum industry, due to their drop size, stability and physicochemical properties. It is one of the chemical-based wellbore cleaning method. Nanoemulsion based spacer fluid is circulated through drill pipe or casing to remove drilling fluid in the wellbore. Nanoemulsion based fluid shows better cleaning due to its smaller droplet size, low IFT, and superior wettability alteration behavior. It is also a better spacer fluid even when OBM is used, thus, ensures wellbore integrity and prevents any damage which can occur to completion equipment. Nanoemulsion is also used as a chemical method of cleaning of oil spills in terrestrial regions, where nanoemulsions cleans spilt oil by dissolving the oil and

© The Author(s), under exclusive license to Springer Nature Switzerland AG 2022   47
N. Saxena et al., *Nano Emulsions in Enhanced Oil Recovery*,
SpringerBriefs in Petroleum Geoscience & Engineering,
https://doi.org/10.1007/978-3-031-06689-4

extracting from contaminated sand. Nanoemulsions can also be used as a chemical EOR method for improving the oil recovery.

The use of nanoemulsion as an efficient EOR fluid can be applied to recover the oil trapped in the pores of the reservoir after primary and secondary oil recovery stages. The injection of nanoemulsion, which have smaller dispersed oil phase, have better oil recovery efficiency in comparison to the conventional O/W emulsion. The physicochemical properties of nanoemulsions such as ultralow IFT with the reservoir fluids, miscibility with the trapped crude oil, wettability alteration efficiency, and higher viscosity with respect to the injection water, causes the change in the reservoir rock and fluid properties. The ultralow IFT of the nanoemulsion causes for the ease of injection of nanoemulsion into the narrow pores of the reservoir, as well as effective miscibility with the crude oil. The solubilization of oil in nanoemulsion changes its properties, such as viscosity, leading it to production from the pores of the reservoir. The injection of nanoemulsion also causes the wettability alteration which leads to detachment of crude oil into the injected nanoemulsion. Nanoemulsions have favorable mobility in reservoir due to its higher viscosity that aids in the recovery of the residual crude oil. Higher viscosity of nanoemulsions leads to its better volumetric sweep, whereas, IFT reduction and wettability alteration leads to improvement in microscopic efficiency, thus, leading to higher oil recovery by application of nanoemulsion.

Nanoemulsion can be potentially applied in most reservoirs with low to medium gravity oil. However, field application of nanoemulsion as EOR agent has been limited, although extensively tested at laboratory scale. The filed application of nanoemulsion will also require significant investment based on requirement of water and oleic phase as well as depending upon the method of formulation of nanoemulsion near the injection site. Tuning of nanoemulsion components and composition to achieve stability in the pores of the reservoir and desired properties for the specific reservoir properties such as formation salinity, formation temperature and formation fluid is required for the success of the nanoemulsion flooding project.